THE
COSMOS
A Beginner's Guide

BBC
BOOKS

THE COSMOS
A Beginner's Guide

HART-DAVIS

PAUL BADER

This book is published to accompany the
television series *The Cosmos –
A Beginner's Guide*

1 3 5 7 9 10 8 6 4 2

Published in 2007 by BBC Books,
an imprint of Ebury Publishing.
A Random House Group Company.

The Random House Group Limited
Reg. No. 954009

Addresses for companies within the Random
House Group can be found at
www.randomhouse.co.uk

A CIP catalogue record for this book is
available from the British Library.

ISBN 978 1 846 07212 3

The Random House Group Limited makes
every effort to ensure that the papers used in
our books are made from trees that have been
legally sourced from well-managed and
credibly certified forests. Our paper
procurement policy can be found at
www.randomhouse.co.uk

Commissioning editor: Martin Redfern
Production controller: Ken McKay
Jacket designer: David Eldridge

tall tree

Created by Tall Tree Ltd
Managing editor: David John
Designer: Ben Ruocco

Consultant: Professor Paul Murdin
Colour Reproduction by Dot Gradations Ltd
Printed and bound by
→ Firmengruppe APPL, aprinta druck,
Wemding, Germany

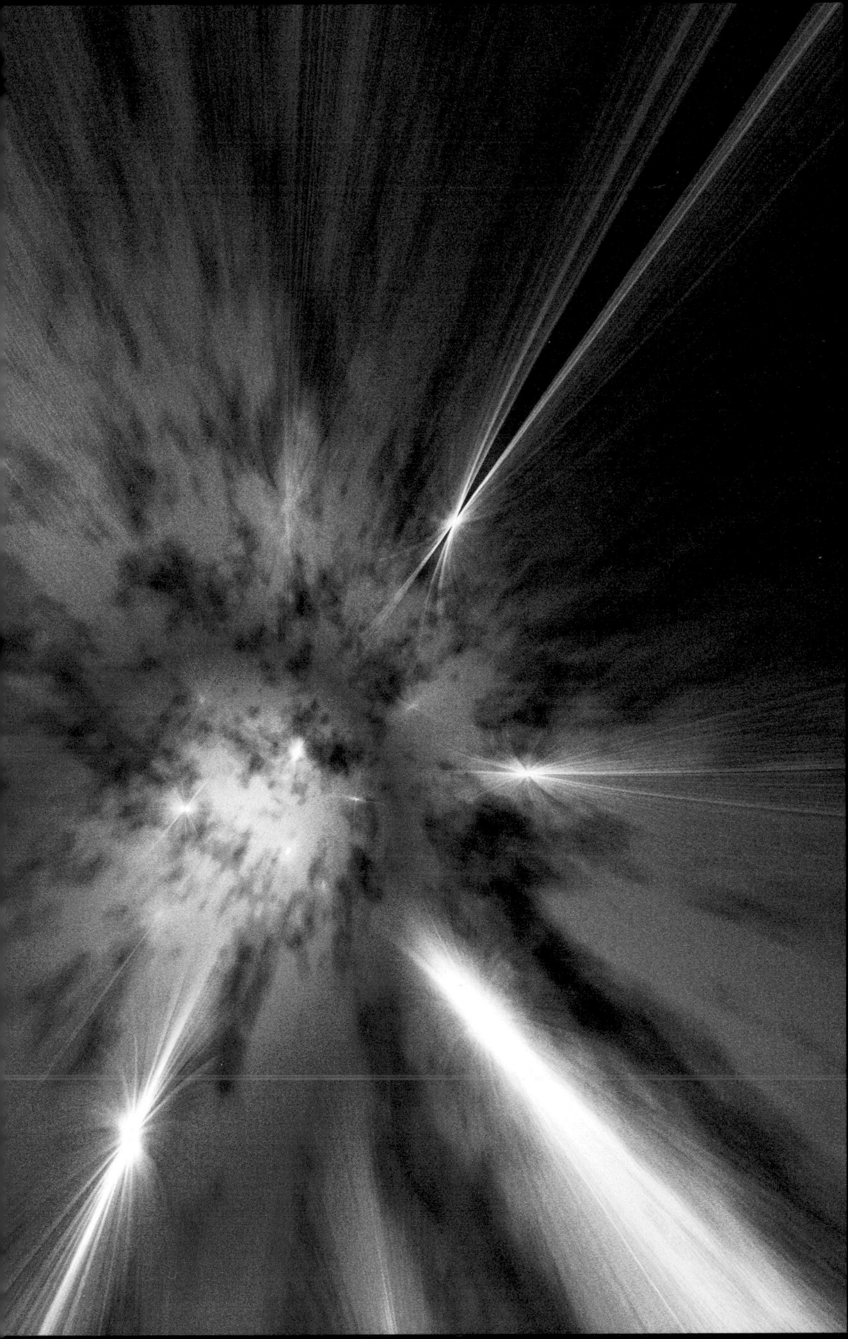

INTRODUCTION

'Cosmologists are often in error, but never in doubt.' This was how the Russian Nobel Laureate Lev Landau joked about those who study the universe. Others have gone further: 'There is speculation, there is fantasy, and then there is cosmology.' With ideas based on rather little hard evidence about the size, shape or age of the universe, or where it came from, it is hardly surprising that the early cosmologists were not thought of as proper scientists. But how that has changed. Although we have still not been able to visit another galaxy or even another star, those who study the cosmos today have discovered ways of using amazing science on even the most distant parts of the universe.

Some of this science is done by astronomers looking out into space, using brilliant telescopes on mountains or on spacecraft to capture light that has travelled to us from way back in time. Other scientists do experiments here on Earth, some of them attempting to recreate the conditions that existed fractions of a second after the Big Bang; others still are rather ambitiously building their own universes inside supercomputers. Practical astrophysics has come of age.

In writing this book, which follows the television series broadcast on BBC2, we went to meet these scientific explorers of the cosmos, and saw the spectacular machines they use. Neither of us is a cosmologist, and many of the ideas that drive space exploration are new to us, so we hope that our struggles to understand and simplify them will prove interesting to our readers. As well as reporting back from the places visited, Adam has attempted to demonstrate some of the ideas in a rather practical way, and we include some of his experiments, with sketches and pages from his notebooks.

In making the series we have travelled to Chile to see the world's largest optical telescope; we have climbed an old volcano in the Canary Islands to visit the planet-hunting SuperWASP; we have been lowered 100 metres underground near Geneva to see the Large

Hadron Collider, and we have checked out the vast new radio telescope being built in California to listen for messages from aliens.

We have been consistently amazed by two things: the extraordinary advances in technology and knowledge that have been achieved in the last 10 or 15 years, and the extraordinary enthusiasm and confidence of the astronomers, the astrophysicists, the engineers and the cosmologists. These people are actively tackling the really big questions: How did the universe begin? What is it made of? Is there any other life in the cosmos? They have unshakeable faith in their work and in the future. They are sure they will make the elusive 'God particle', find more planets, and make contact with another intelligent civilization in our galaxy.

This book has six chapters, corresponding to the six television programmes: How to build a universe, Seeing the universe, Space exploration, Other worlds, Violent universe, and Are we alone? Although at first glance some of these topics are dissimilar, we found one underlying idea cropping up again and again: are there other intelligent creatures in the universe? This is explicit for the SETI people, but it also seemed to be a fundamental question for the planet-hunters, the space explorers and even perhaps the universe-builders. They all want to know the answer to Life, the Universe, and Everything, but in particular they want to know about intelligence; that is why you will find the question coming up in every chapter.

For us, writing this book has been a wonderful adventure; it has been a privilege to meet these brilliant people and to hear what they have to say. For Adam the most satisfying experience was taking photographs of the laser-guide star at the Paranal Observatory in the Chilean desert (see page 52). For Paul it was looking up at the night sky and seeing for the first time our neighbouring galaxies, the Magellanic Clouds (see page 40).

Adam Hart-Davis
Paul Bader

February 2007

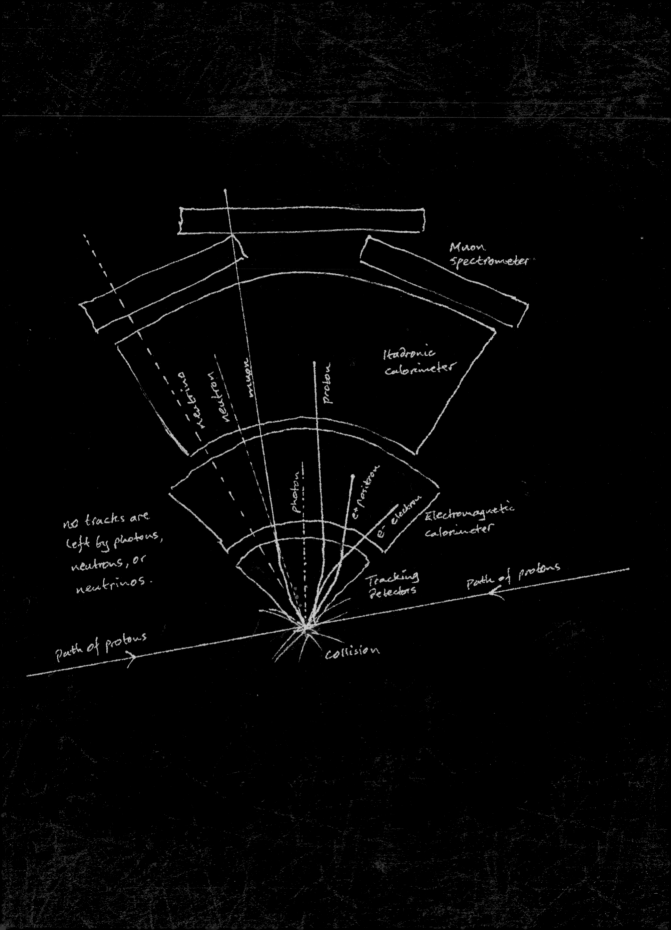

Muon
Spectrometer

Hadronic
calorimeter

neutrino

neutron

muon

Proton

Photon

e⁺ positron

e⁻ electron

Electromagnetic
calorimeter

no tracks are
left by photons,
neutrons, or
neutrinos.

Tracking
Detectors

Path of protons

Path of protons

Collision

 1 / HOW TO BUILD A UNIVERSE

Just imagine if you could build your own universe. In the past the very notion would have been thought impertinent – by those who believe that creating universes is God's job – or just plain absurd. But today, incredible as it seems, there are scientists not only thinking about building a universe, but actually starting to do it. Virtual universes full of billions of galaxies and accurate in every astrophysical detail have been developed inside supercomputers. And soon, deep beneath the French-Swiss border, a machine will be switched on that for the first time will unleash temperatures and forces that have not existed since the Big Bang – the event that created our universe. When this happens it will take our understanding of the universe to a new level.

You might well ask why anyone would go to the trouble and expense of trying to build a universe. The short answer is: to find out how it's done. We cannot rewind time to see how the universe began, so trying to make a DIY universe is the practical alternative. If we get the method and the ingredients right, then we should end up with a universe that looks like our own.

'A continual miracle is needed to prevent the sun and the fixt stars from rushing together through gravity'

Isaac Newton

DATA FILE: ANCIENT COSMOLOGY

The question of how the universe began was not one asked by all ancient cultures. The ancient Chinese believed that everything, and all events, in the universe are going through endless cycles. Hindus also had a complex cyclic system, though their cycles were incredibly long, lasting up to trillions of years (1 trillion = 1 million million). The story of a universe created by God is mainly found in the Judeo-Christian tradition.

For centuries we have looked out into space and speculated, inventing hypotheses to explain how the universe began. The science of understanding the universe is called cosmology. It started as a theoretical and rather a fringe discipline, not much respected by most scientists. Recently, however, cosmology has transformed itself into an experimental enterprise that conducts some of the largest and most expensive investigations in the whole of science, using some of the most complex machines in the world.

Although we now use high-energy physics machines to probe the universe, the great revelation that launched modern cosmology was made in a series of fairly simple telescopic observations in the 1920s. This was an exciting time in physics, because the scientific community was getting to grips with the two papers on relativity theory published by the great German-born physicist Albert Einstein in 1905 and 1915. The second of these introduced the concept of space-time and set out an entirely new idea about the nature of gravity. Since the time of Isaac Newton, the English mathematician, gravity had been thought of as an attraction, a mysterious force that pulled two things with mass together. It is what holds all of us down on Earth, and keeps Earth in orbit around the Sun. But although Newton had described gravity, and given us some handy mathematics to go with it, there remained two great mysteries.

First, if there was all this attraction, with everything pulling on everything else, why didn't the universe collapse together in a big heap? In 1692, an English parson named Richard Bentley wrote to Newton pointing this out, and noting that we don't see evidence of stars crashing into each other. Newton had to think of something. One option was to say that the universe is infinite; then you might be able to imagine an equal pull in all directions, and everything would remain stable. But Newton did not like the idea of an infinite universe, and instead made up what in truth was an excuse, saying that everything had been set up so that there was a perfect balance.

The second problem with Newton's gravity was that he couldn't say how it worked. How could gravity work invisibly, and over great distances, even from one side of the universe to the other? It seemed like a conjuring trick, and Einstein's achievement was to come up with a radical new explanation for gravity that was as applicable to apples falling to the ground as it was to planets orbiting the Sun. By bringing together space – the normal three-dimensional world we experience – and time, Einstein could finally explain gravity as something that happens when massive things distort what has come to be known as space-time. The great thing is that this happens locally, but the distortion can affect distant objects. Think of two people on a soft bed: the heavy one will create a dip, and if the dip is big enough the lighter one will roll down into it. Einstein's space-time is the mattress.

RELATIVITY TRIUMPHS

What separated Einstein from previous theorists who speculated about the universe is that he made predictions that could be tested. For instance he predicted that really massive objects, such as the Sun, would distort space-time so much that you'd notice light curving as it passed by. In 1919 British astronomer Arthur Eddington proved Einstein right when he showed that light from more distant stars was indeed curved as it passed the Sun. This was tricky because by definition when you see the Sun – in daytime – you usually cannot see the stars. So Eddington took advantage of a rare total eclipse of the Sun, which allowed him to see Sun and stars in the sky at the same time. Sure enough the gravity of our local star caused light to bend because the stars' positions around the Sun appeared to have shifted slightly in the sky. It was the first experimental confirmation of Einstein's ideas, but by no means the last. Testing his ideas is still one of the things that drives experimental cosmology 100 years after they first shook the world. In the 1920s Einstein was already a superstar. But a new figure was waiting in the wings whose discoveries would shock the world – and cause Einstein to admit his greatest mistake. That man was an American astronomer named Edwin Hubble.

BIOG FILE:
ISAAC NEWTON

Born in Lincolnshire, England, in 1642, Newton was a mathematician, physicist, astronomer and alchemist. His *Principia Mathematica*, published in 1687, described gravitation and the three laws of motion. Newton was the first to show that the motion of objects on Earth and of celestial bodies are governed by the same natural laws. This discovery was integral to the scientific revolution of the seventeenth century, the acceptance of the idea that Earth goes around the Sun, and of the notion that scientific enquiry can reveal the mysteries of nature.

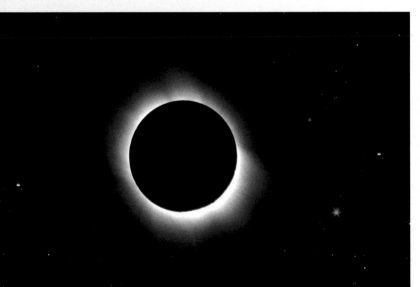

During the few minutes of a total solar eclipse, Arthur Eddington measured the apparent position of stars close to the Sun. They weren't in the positions conventional astronomy predicted. The only explanation was that the path of light from the star, and the space-time through which the light passed, had been warped by the gravity of the Sun, as Einstein had suggested.

BIOG FILE:
ALBERT EINSTEIN

Albert Einstein was born in Ulm, Germany, in 1879 and showed mathematical ability from an early age. He won a Nobel prize for his work on the properties of light, but it was his theories of special relativity that made him the most famous scientist of the twentieth century. These showed, among other things, that nothing moves faster than the speed of light, that this speed is always constant, and that objects become more massive as they move faster. He also found that mass is equivalent to energy. In what has become the most famous equation in the world, Einstein stated that energy (E) was equal to mass (m) times the speed of light (c), squared: $E = mc^2$. After Hitler came to power in 1933 Einstein renounced his German citizenship and continued his work in the US, where he died in 1955.

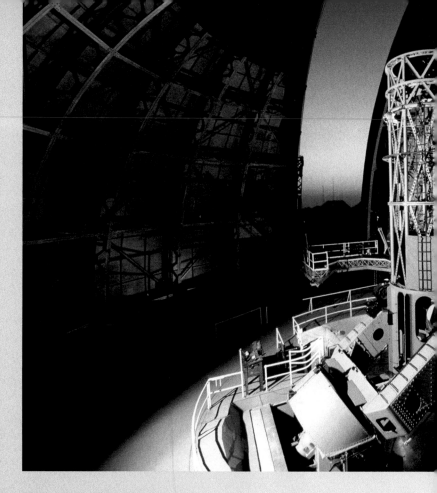

THE SIZE OF THE UNIVERSE

At 1742 metres (5715 feet) above sea level, the Mount Wilson Observatory in Pasadena, Southern California, usually sits above the cloud that can envelop the region. The clear skies made it the ideal site for the great '100-inch' (2.5-metre) Hooker Telescope, the largest in the world from 1917 to 1948.

What was the Hooker for? Although magnification is useful, the key benefit of a large mirror is that it can gather a huge amount of light. This helps to reveal fainter stars and something that particularly interested astronomers of the 1920s: the nebulae. In Latin, nebula means 'mist'. Nebulae are faint, cloudy objects that range from huge (some of them occupying as much of the sky as the disc of the Moon) to so small in appearance that before the use of big telescopes they were originally confused with stars. With a big enough telescope some of the nebulae appeared to have a spiral structure. But what were they?

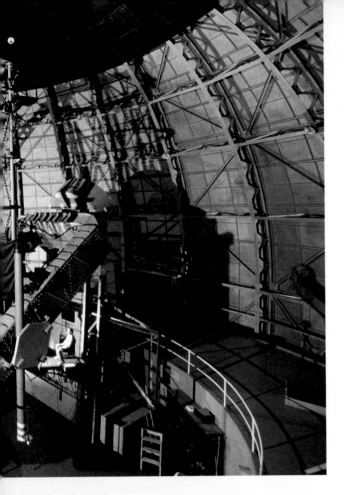

Two of the world's largest telescopes were built on Mount Wilson. The first, the 1.5-metre (60-inch), was finished in 1908, surviving the San Francisco earthquake of 1906. The '100-inch' (2.5-metre) Hooker (left) was built soon after.

In the debate Shapley argued that there was only one galaxy, that it was large (over 200,000 light years across), and that our Sun was far from the centre. Spiral nebulae were distant clouds of gas or dust in this single galaxy-universe. Curtis took the opposite view. Our galaxy was smaller (60,000 light years across), with the Sun near its centre, but crucially he argued that the spiral nebulae were *island universes* – separate galaxies like the Milky Way. In his model, the universe consisted of many island universes.

In previous years, the debate would have been accompanied by a glass of wine. But in America the era of prohibition had just begun, so alcohol was out. Each of the protagonists read his paper, and then responded to the other. There was no resolution, and there couldn't have been one that year. Both Shapley and Curtis based their arguments on the data available, their own and those of others. Had the evidence been conclusive, there would never have been a debate. Nevertheless it seems remarkable that well into the twentieth century the leading astronomers of the day disagreed about the size of the galaxy; whether our Sun was at the centre of it; whether the galaxy was the same thing as the universe; and whether the cloudy spiral nebulae were merely odd features of Milky Way, or complete separate galaxies – island universes – in their own right. In hindsight, it seems as though some of these arguments were still fuelled by a belief that humanity must be at the centre of all things. Copernicus may have removed Earth from the centre of the solar system, but we still wanted to

Although it was well known by the early twentieth century that our solar system is part of a galaxy, the Milky Way, the structure of the universe was hardly known. Most astronomers thought the Milky Way *was* the universe, and that everything you could see in the night sky formed part of the same system. But the spiral nebulae worried some observers. They began to suggest that these cloudy objects were not part of the Milky Way at all, but were 'island universes' in their own right, and that the Milky Way was itself an island universe. This was part of a discussion about the size of the universe that culminated in the 'Great Debate' between the American astronomers Harlow Shapley of Mount Wilson and Heber D Curtis of Lick Observatory, California, during the National Academy of Sciences meeting in Washington DC on 26 April 1920.

13

BIOG FILE: EDWIN HUBBLE

Hubble was born in Missouri in 1889. He studied science in Chicago, and then went to Oxford where he switched to law – the result of a promise to his dying father. He was an impressive, athletic figure who played basketball, and was at one time offered the chance to become a professional boxer. Hubble returned to the US to teach Spanish before returning to astronomy. He was eventually offered a post at Mount Wilson, but service in the First World War kept him away until 1919, the year before the 'Great Debate' about the scale of the universe. He later campaigned for astronomers to be eligible for the Nobel Prize in Physics, something that happened only after his death in 1953.

sit at the centre of at least some universes. There was, however, agreement on one thing: Curtis's call for more evidence. And the man who was to provide the evidence that would change our view of the universe for ever was already at work.

HUBBLE'S DISCOVERY

Edwin Hubble was a big character. His achievements were extraordinary though not everyone thinks of him fondly, claiming that he aggressively promoted his own contribution, forcing others into the shade. But his is the name that lives on, especially in the form of the Hubble Space Telescope.

At Mount Wilson, using the Hooker Telescope, Hubble tackled the problem of spiral nebulae. Although described as dusty clouds, it was already apparent that they contained things that could be stars. Hubble's breakthrough came when he noticed the presence of a particular sort of star in the Andromeda Nebula that he could use to measure distance. The idea is fairly simple. The further away a star is, the fainter it looks. Measure the brightness, and you can estimate its distance. The problem is that there are many types of star, so you have no idea how bright it ought to look at a particular distance. Variable stars change in brightness, and had been known about for hundreds of years. A particular sort of variable star, the Cepheid variables, have a unique property. Their brightness and the speed with which they get brighter and dimmer are related. By noting the rate of varying brightness, you can work out how bright the star actually is. Then you can tell how bright it ought to look at any distance from Earth.

Hubble's discovery of a Cepheid variable star in the Andromeda Nebula meant that he could measure how far away it was – the first time this had been done. He did the sums, and the result left no room for doubt. Andromeda was 2.5 million light years away – so far beyond even Shapley's estimate of the size of the Milky Way that it clearly existed way outside our own galaxy, and was an island universe, or galaxy, in its own right. This was particularly shocking because the Andromeda Nebula – now called the Andromeda

The Andromeda Galaxy is here seen in natural light. Recent observations by the Spitzer Space Telescope reveal that it contains one trillion (10^{12}) stars, greatly exceeding the number of stars in our own galaxy.

'...we are reaching into space, farther and farther, until, with the faintest nebulae that can be detected ... we arrive at the frontier of the known universe'

Edwin Hubble

Galaxy – is the nearest great spiral galaxy to Earth. With this one simple observation, Hubble had dramatically increased the size of the known universe.

One of Hubble's great strengths was a seemingly inexhaustible capacity for observation, and he soon set the distances for many other galaxies. It was now clear, thanks to Hubble, that the universe was far, far larger than anyone had thought, and consisted of many, many galaxies of which the Milky Way was just one. The idea was not entirely new, having been suggested previously by others including the Estonian astronomer Ernst Öpik, though it was Hubble who nailed the proof. But he wasn't done yet.

Hubble now wanted to know more about these other galaxies. First he conducted a survey of the skies. Where could galaxies be found? The answer: everywhere. He found galaxies in every direction, and right to the limits of the Hooker Telescope. The universe, it seems, is filled with galaxies. He looked for more detailed information about individual galaxies, especially how they were moving in the universe, and what they are made of. Their light, reaching Earth over great distances, contains all that information, and Hubble had the equipment to decode it.

DATA FILE: CEPHEID VARIABLE STARS

The type of variable star known as a Cepheid varies in brightness because it pulsates, or regularly expands and contracts, as it nears the end of its life. As it gets bigger, it cools and emits less light. As it gets smaller, it becomes hotter and emits more light. This correlation was noticed and given a precise mathematical form by the American astronomer Henrietta Leavitt in 1908. Cepheids are an accurate guide to measuring distance.

DATA FILE:
DOPPLER EFFECT

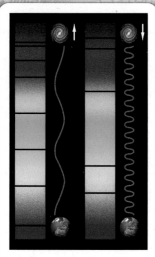

When an ambulance siren goes past, the note you hear changes. It sounds higher when it is approaching, lower when it is going away. How the note sounds depends on the number of sound waves reaching your ear every second. This was discovered by the Austrian physicist Christian Doppler in 1842. He predicted that the same effect happens with light – if a light source is moving towards you or away from you the light waves arrive at a different rate. In this case though, you would see a redder or a bluer colour. Hubble was sure that sodium light was the same yellow in far galaxies as it is here, but the Doppler effect meant that it would be shifted in the spectrum towards red or blue if the galaxy were moving away or towards us.

You may have noticed that the bright yellow streetlights used in many parts of the world are exactly the same colour yellow. This is because their bulbs contain sodium vapour, and when electricity is used to jiggle it about, sodium emits a very characteristic colour of light. If sodium is there, you get this very precise yellow. This is useful to know because by examining sunlight, done by producing a very wide and spread-out rainbow, you can see whether that particular yellow is present, and therefore whether the Sun has sodium in it. We explain more about what happens when light is split in chapter 2.

The rainbow, which looks continuous to our eyes, really consists of many coloured lines that come from the elements inside the Sun emitting their very particular colours. It is worth emphasizing again that the light from a chemical element is always a particular colour, so that if you find that line in the spectrum you can be pretty sure what is causing it. This technique was used to work out what elements are burning in the Sun, and in stars – proving that they, too, are suns. Spread the colours wide enough and the so-called Fraunhofer lines appear, each associated with a chemical element (see page 60).

Hubble wanted to learn about the elements in the galaxies he was looking at, but he mainly planned to use their spectra to measure how fast they were moving, and in what direction. The idea for doing this wasn't his, but once again it was Hubble who put it into practice and produced such overwhelming evidence that no one could doubt him. Having emphasized that yellow sodium light is always exactly the same yellow here on Earth, it turns out that sodium light in other galaxies is a slightly different yellow, either slightly redder or slightly bluer. This is not because the chemistry of those galaxies is different from our own, but because the galaxies are moving – an effect that causes the colour of known elements to change slightly (see box).

Hubble began by measuring the blue-shift or red-shift of galaxies. Our closest galactic neighbour, the Andromeda Galaxy, has a blue shift. It is coming towards us. This in itself is a remarkable finding, given that before Hubble's work a couple of years earlier, most people didn't realize that Andromeda was a galaxy. In recent years it has been suggested that it might one day collide with the Milky Way. Then he looked further out. Soon a bizarre and disturbing pattern began to emerge. Beyond a certain distance, there was no blue-shift. Galaxy after galaxy were red-shifted by different amounts, indicating that they were travelling at different speeds – but all the distant galaxies were red-shifted. Hubble wasn't afraid to reach the logical conclusion, no matter how controversial – and this conclusion was certainly that. If all the distant galaxies were red-shifted, that must mean they are moving away from us.

But how can all distant galaxies be moving away from us? Hubble had the means to answer at hand. Remember, he had already measured the distances to galaxies using the Cepheid variable stars. Now he matched the distances to the red-shifts, and what he found was the sort of striking

match scientists dream of. The graph gave a straight line: the further away the galaxy, the greater the red-shift. In other words, the further from Earth a galaxy is, the faster it is moving away from us. At first this appears to imply that because everything is retreating from Earth, then we must somehow be at the centre. But Hubble had the sense to realize that the universe would look like this from wherever you were because the whole thing is expanding.

THE EXPANDING UNIVERSE

When Edwin Hubble announced his results, in a series of papers published between 1929 and 1936, not everyone was convinced – including Albert Einstein. Einstein had long ago known that his equations could describe a gently expanding universe, but he thought the result must be wrong. Like many others, Einstein was convinced that the universe was unchanging and had always been pretty much as it now appears. So he fudged his equation, adding an extra term (the 'cosmological

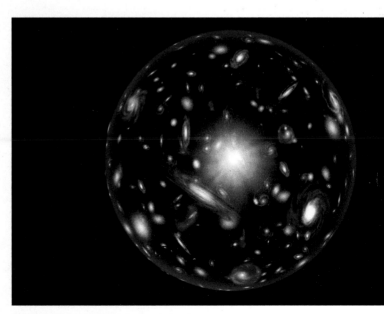

This picture tries to show galaxies in an expanding universe that is limited in size but unbounded. The picture is two dimensional but shows a three dimensional object representing a four dimensional universe. The yellow glow in the centre represents the origin of the universe in the Big Bang. The galaxies are shown on the surface of an inflating balloon that originated at its centre. The balloon's surface has no boundaries but is finite in area. The galaxies are moving away from each other on the growing surface, not from a central point.

BIOG FILE: GEORGES-HENRI LEMAÎTRE

Georges-Henri Lemaître was a Belgian priest who would eventually head the pope's science academy. Lemaître was born in 1894 and received the Military Cross for bravery in the First World War before taking holy orders. He was unusual in combining religion and science. Although he seems an odd figure today, his science credentials were rather good: he studied at Cambridge under Arthur Eddington, and then under Harlow Shapley in Massachusetts. He had also mastered Einstein's papers, and suggested that they could be extended to describe an expanding universe, and therefore what he called the 'hypothesis of the primeval atom' – the Big Bang. Lemaître died in 1966 shortly after learning of the discovery of cosmic microwave background radiation, proof of his theory about the birth of the universe.

constant') that acted against the expansion and kept the universe on an even keel. But now Hubble had evidence suggesting that Einstein had originally been on the right track. The implications of an expanding universe caused Einstein a major headache. If the universe is currently expanding, that must mean that in the past it was much smaller than it is now. But how much smaller, and when did the expansion start? This could be worked out. By assuming that the universe expanded at a constant rate, and projecting backwards from the present day, everything gets smaller and smaller until it reaches a single point about 14 billion years ago. If Einstein was worried about this startling possibility – that the universe started in a single point – you can imagine that many others were worried, too. Just because Hubble produced evidence that the universe was expanding didn't mean that it had always been expanding, they argued. One of those who preferred the idea of a 'steady state' universe was the English astronomer Fred Hoyle. He and others couldn't deny the evidence, but suggested that the expansion we see is compensated by contraction happening somewhere else – what he called 'continuous creation' – and the overall effect is a steady state. Attempting to dismiss it, Hoyle ridiculed the idea that the universe started in a 'Big Bang'. The name stuck better than Hoyle's own ideas.

If you had to imagine the sort of scientist who might help cement the idea of the Big Bang, Father Georges-Henri Lemaître probably wouldn't be what you had in mind. This Jesuit priest first published his ideas in 1927, before Hubble's results, but no one took much notice. As a religious man, the idea of a universe made in one day must have been appealing, but Lemaître's ideas were based on sound mathematics. He stuck with them and in 1933 found himself at a conference in California where Einstein himself was in the audience. At this event the penny finally dropped for Einstein, who realized that Lemaître's ideas and Hubble's evidence were irresistible, and that his unnecessary invention of the 'cosmological constant' – to prevent his own equations from describing an expanding universe – was a huge mistake.

The expanding universe quickly became the accepted model. It provided a new way to think about what had been seen through the Hooker Telescope. The red-shift of distant galaxies occurs because space itself

This is an imaginary depiction of the Big Bang. Scientists can use the laws of physics to trace the history of the universe back to a trillionth of a second after the Big Bang, but they still don't know what happened in the very moment the universe began, because physics as we know it did not apply.

is expanding, and as light passes through it, its wavelength gets longer, making it look redder to us. The challenge for science is now to understand how you get from the Big Bang to a universe that is at least 14 billion light years across, and filled with galaxies. What were the ingredients? How do you build a universe?

Today, a two-pronged investigation is under way. Physicists are attempting to use enormous energies to create conditions similar to those that existed instants after the Big Bang – but this time in a giant laboratory. And astronomers are peering deeper and deeper into space at objects whose light is reaching us from an age when the universe was much younger. Will it be possible to see as far back as the Big Bang itself, or at least to see something that would be certain evidence for it?

RECREATING THE EARLY UNIVERSE

CERN, near Geneva in Switzerland, is the largest particle physics laboratory in the world. Experiments here involve smashing particles together to reveal not only what they are made of, but what is at the heart of matter itself. When at the end of 2007 a new machine at CERN is turned on, it will create inside itself a tiny fireball that will have temperatures not seen since about half a second after the Big Bang. In this tiny inferno physicists hope to find signs of what conditions were like at the start of our universe, when matter as we know it, even the particles inside the atoms, did not exist. Instead scientists hope to detect strange things that might explain why the cosmos is the way it is.

DATA FILE: THE BIG BANG

From Hubble's telescopic results came the stunning idea that our universe began in an instant, with everything, including matter, space and time, emerging from a single point about 13.7 billion years ago. The new-born universe was infinitely hot and dense, but within a trillionth of a second particles appeared, created out of pure energy. As the temperature dropped, gravity separated from electromagnetism and the nuclear forces, releasing huge amounts of energy that caused the universe to balloon faster than the speed of light. In under a second, it grew from the size of a pinhead to the size of a galaxy. After 300 million years atoms formed and material began to concentrate into narrow filaments with huge voids between them.

CERN's Large Hadron Collider

Walking around CERN, you soon realize that getting to the smallest things in the universe – subatomic particles – requires one of the largest and most complex machines on Earth: the Large Hadron Collider (LHC). In essence the machine is a simple device – a long tube in which protons (a proton is the nucleus of a hydrogen atom) are sped up to 99.99999 per cent of the speed of light, and then smacked into each other. The resulting fireball sprays particles off in all directions, then a machine known as the Atlas detector has to find out what happened. If it finds evidence of the Higgs particle – sometimes called the 'God particle' – we may finally know why things have mass. If some of the collisions

seem to defy the laws of symmetry, we may get a handle on dark matter, which makes up a quarter of the universe but has never been seen. Solve mass and dark matter, and it would be Nobel Prizes all round. But getting there needs the LHC to do what no machine has done before.

The LHC tunnel is 100 metres (330 feet) underground. It was made for CERN's previous experiment, which used electrons, and this is part of the challenge. Ideally, a much bigger ring is needed for protons, which are 2000 times more massive. As they get faster and faster in the ring, it is harder and harder to keep them going round. To do so, magnets are needed, and

The tube in which the LHC accelerates protons is 27 km (17 miles) long, circular and buried 100 metres (330 feet) below the French-Swiss border. You can cycle round the whole thing and no one asks for your passport.

The interior of the LHC's tube contains the two smaller tubes in which protons are sped clockwise in one, anticlockwise in the other.

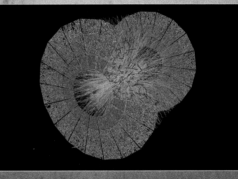

This is a simulated computer image of an ion particle collision inside the LHC. It produced a type of plasma that existed microseconds after the Big Bang.

using over 1000 magnets (1,232 to be exact) it is just about possible. These magnets themselves are superconducting, which means they have to be very cold – just 1.9 degrees above absolute zero: -273°C (-459°F), the lowest temperature possible. There are so many miles of wire filaments inside the magnets that they could stretch from Earth to the Sun. Five times.

But they are worth it, because once these protons are in motion, the magnets will give them far more energy than was possible in the electron experiments, and extreme energies are needed to get close to Big Bang conditions. The

need a leak-proof tube 27 km (17 miles) long. The vacuum is so good it can't be measured – if a probe were to do so it would introduce more molecules than there are inside the tube already. For most of the 27 km (17 miles) there are two tubes, one for protons heading in a clockwise direction, the other for protons heading anticlockwise, and they are brought together only where they are meant to collide, which is in each of the experiments like the Atlas detector (see next page). At full speed, protons make it round the loop 11,000 times a second in both

The purpose of all this hardware is to get as many collisions as possible, because most of them are not going to be interesting. Given the billions of particles involved, the hit rate is small – despite all their efforts at concentrating the beam of protons, CERN's engineers end up with mostly empty space. The magnets round the tunnel accelerate and focus the beam, making bunches of protons a few centimetres long, and thinner than a human hair. Each bunch contains 100 billion protons. Two bunches, one from the clockwise tube, the other from the anticlockwise tube, head straight at each other. But most pass on without colliding. Even with 200 billion protons involved, there are expected to be only 10 to 20 collisions. But this isn't so bad when you consider that 40 million bunches pass through in each direction every second. So when all is going well, there could be 600 million collisions a second – any of which might be the one the scientists are looking for. The job of the LHC is to supply colliding protons with enough energy to create mysterious particles like the Higgs. Finding the particles happens in the detectors, of which Atlas is the largest.

ATLAS

If the scale of the LHC can only be appreciated by walking round it and seeing the thousands of magnets and kilometres of cables and tubing, Atlas can be taken in at a glance, and it is absolutely vast. It is arguably the most complex machine on Earth. It is 46 metres (150 feet) long and 25 metres (82 feet) high, and its job is to wind back time. Explained simply, Atlas will work something liked this. Protons will enter at each end through a tube 28 mm (1.1 inch) in diameter. They are in bunches, and going very fast. Six hundred million times a second, there is a collision between a pair of them. The protons are destroyed, but out of the fireball comes a spray of new particles shooting in all directions. Atlas has to track these and identify what they are. The first problem is that no one detector can do the job, so Atlas consists of layer upon layer of different detectors, set up so that the ones nearest the centre will pick up those particles that don't last very long, and the ones at the outside are sensitive to anything that makes it through all the other layers. One consequence of this is the vast amount of hardware involved. Information overload will be one of the big problems for Atlas. With 600 million collisions a second, and hundreds of thousands of bits of data for each

This is the Atlas detector under construction in a vast cavern. See how tiny the engineer working on the gantry looks. Atlas is made of layer upon layer of detectors, all surrounding the point where protons collide. The collisions produce showers of particles, some of which are so short-lived that they will never be seen; instead Atlas will pick up the debris when they disintegrate. From this secondary evidence scientists will know that the original particle existed.

collision, it will be completely impossible to store it all for consideration later. So they don't. Instead each collision 'event' will be analysed automatically by very fast computers as soon as it has happened. But it has to be quick. Of the 600 million events, only between 10 and a 100 are worth recording – the rest are junked. This will still result in a vast amount of data – up to 3 million events a day when the LHC is running as it should. Even so, the really interesting events, ones that might reveal a Higgs particle or dark matter, are incredibly rare: if they find more than one a year they will be doing well.

Atlas seems a world removed from the Hooker telescope used by Hubble on Mount Wilson, and not just because the technology is of a different order. Atlas looks for the unimaginably small; the Hooker looks for the unimaginably large. Yet both are involved in the same quest, to understand how the universe works, and especially how it began. Until now, the rules that governed the very big and the very small have been quite separate. The discoveries at CERN may bring them together, especially if it turns out that tiny particles are responsible for gravity and mass, which determine the movement of stars and galaxies.

INSIDE THE ATLAS DETECTOR

The proton collisions inside Atlas are monitored by layers of detectors, each trying to find out what happens when the protons smash. The innermost layer, of pixel detectors, is grouped tightly around the tube where the collisions take place. Think of these as sensors in a 50-million-pixel digital camera. Beyond those are tracking detectors, which track the directions of particles. Then there are calorimeters to measure the energy of particles, and finally there are muon detectors. Muons are particles that pass through everything else without being altered, so these detectors can be placed around the outside of Atlas where the diameter is now vast.

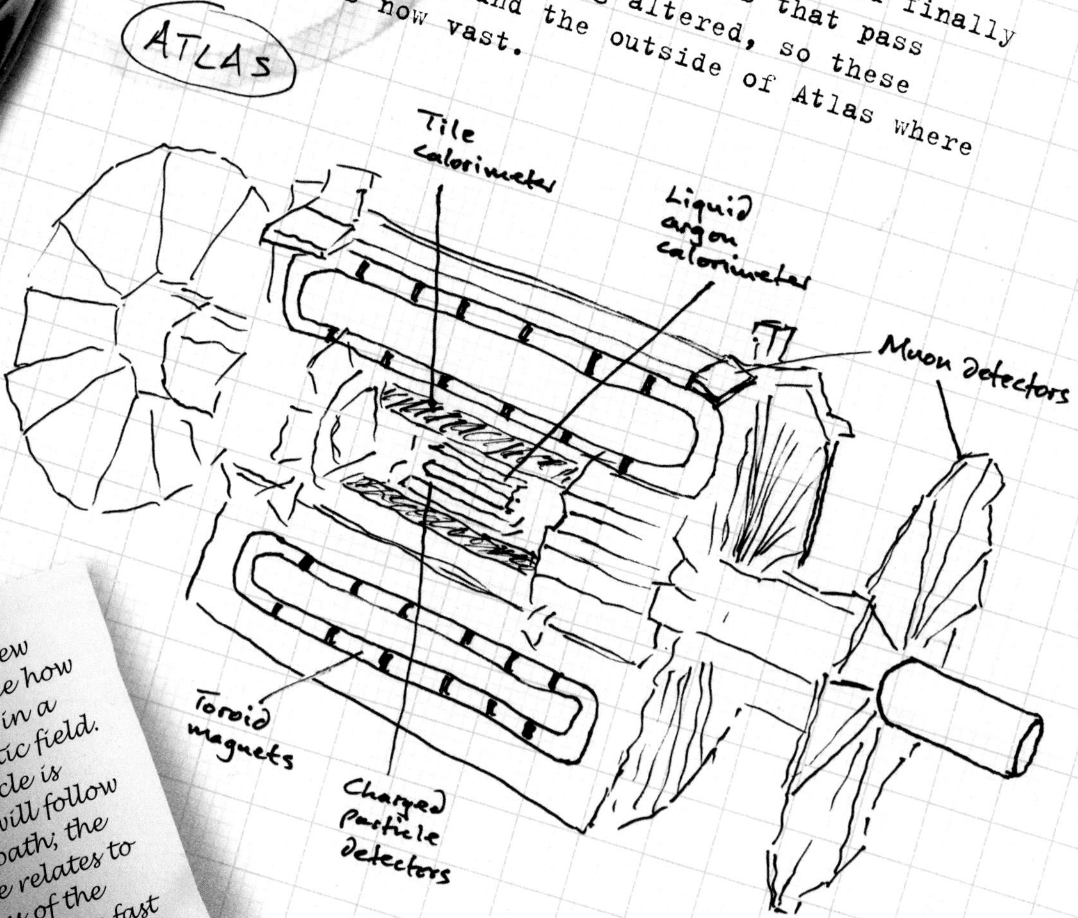

ATLAS

Tile Calorimeter

Liquid argon calorimeter

Muon detectors

Toroid magnets

Charged Particle detectors

One way of identifying new particles is to see how they behave in a strong magnetic field. If a particle is charged, it will follow a curved path; the exact curve relates to the mass of the particle and how fast it is going.

Tracking particles isn't easy. Many are short-lived, or may break down into other particles and only these will be detected. This is true of the elusive Higgs particle. By tracking the particles it breaks into, scientists hope to confirm that Higgs was there.

Scientists using Atlas are trying to wind back time. From a series of tracks left by particles in the detector, is it possible to track them back to the centre and work out what happened in the original collision?

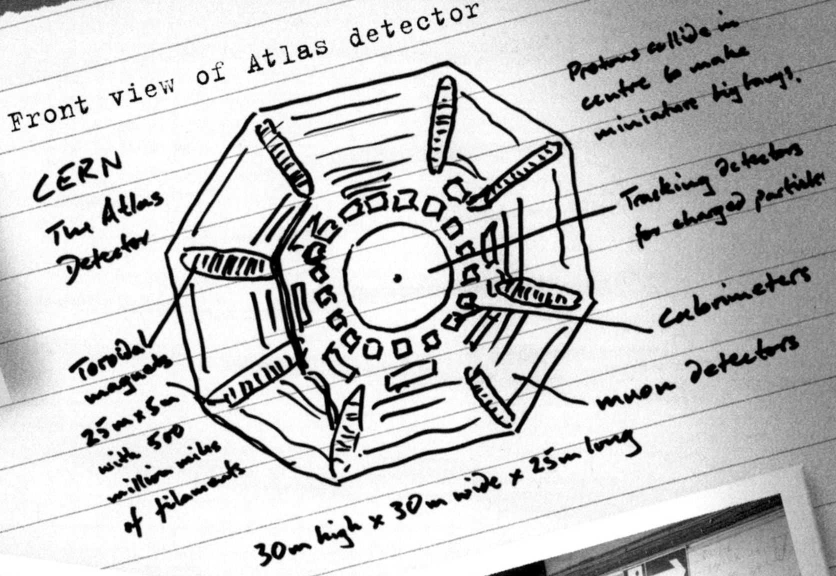

Front view of Atlas detector

CERN
The Atlas Detector

Protons collide in centre to make miniature big bangs!

Tracking detectors for charged particles

Calorimeters

muon detectors

Toroidal magnets
25m x 5m
with 500 million miles of filaments

30m high x 30m wide x 25m long

In our 'snooker detector', a semicircle of white paint is the detector; the snooker balls are the colliding protons. Nothing inside the ring is recorded - just as in Atlas.

After the collision, the 'detector' displays the tracks left by the 'protons' so we can follow them back and work out what happened.

no tracks are left by photons, neutrons, or neutrinos.

photon

e⁻ e⁺

Electromagnetic calorimeter

Tracking detectors

Path of protons

Path of protons

Collision

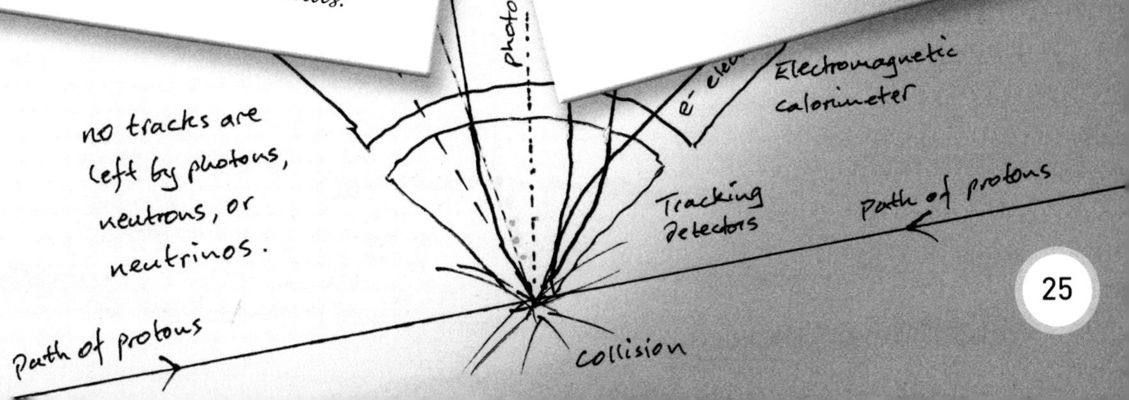

Professor Carlos Frenk has made many universes. He has recently perfected the recipe, and the simulated universe on his monitor has developed into one very similar to our own.

> *'It is not in the nature of cosmologists to be modest'*
>
> Carlos Frenk

Seen here are the evolving stages of a 'successful' virtual universe. Matter concentrates into filament-like structures and eventually into galaxies.

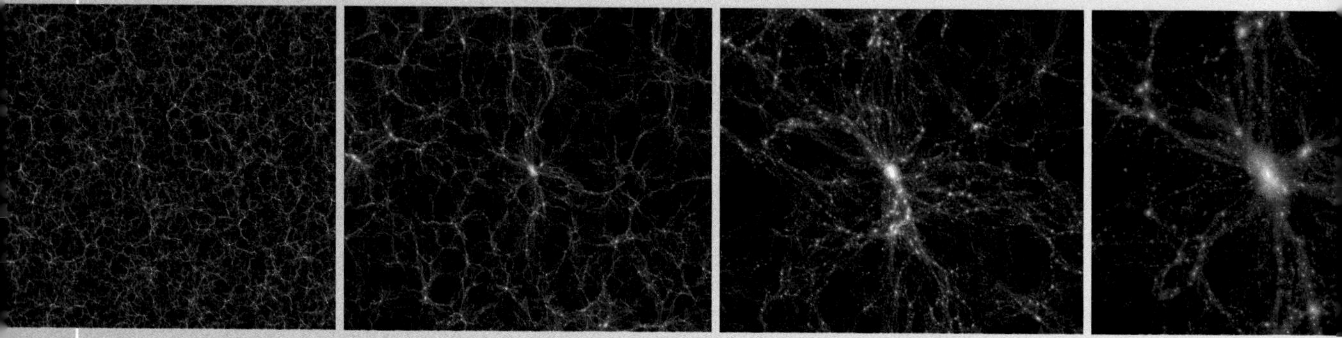

VIRTUAL UNIVERSES

It would be convenient if we could simply watch a movie of the beginning of the universe in order to see which ripples just after the Big Bang led to galaxies, and how the various ingredients worked together. Impossible of course... unless you travel to Durham to meet Professor Carlos Frenk. Not only has Professor Frenk made the movie, he's made the universe, too. In fact he's made loads of them. Recently, he's been getting it right, and his universes look remarkably like the real thing.

Although Professor Frenk's movies are very beautiful, and with a pair of polarizing glasses, they can be enjoyed in 3D on a giant screen in his visualization lab, he points out that what you see is science, not art. It is a simulation, based on the laws of physics. These are known absolutely, and he presumes that they applied at the start of the universe just as they do today. What is not known are the ingredients for a good universe – meaning one that turns out like ours. So Frenk calls what he does 'cosmic cookery'. Select ingredients, pop into a computer, and let it cook.

In truth, the Institute for Computational Cosmology team at Durham do not make complete universes – even their computers couldn't handle that – but start with about 10 billion particles and the aim of making a large cube of universe. Computing power is provided by Cosma, the ICC's cosmology machine, which is one of the most powerful computers in the UK. Today's supercomputers are made of hundreds of processors working in parallel, clocking up a trillion calculations a second at peak. The complete simulation can still take weeks or even months to complete: even this virtual universe is pretty big.

Carlos Frenk has recently completed the largest-ever universe simulation, the Millennium Simulation. It follows the evolution of 20 million galaxies over 2 billion years. It impresses technologists because of the sheer volume of data the simulation handles; and it impresses astronomers because the stunning 3D visuals look very much like the real thing. There are now very big, detailed, digital surveys of the universe, made possible by digital technology in both telescopes and computers, so that where astronomers used to concentrate on a few interesting objects, these surveys can take in vast tracts of the sky. It is the 'real' universe recorded in these surveys that the ICC team compares with their simulated universe. The match is remarkably close.

But Carlos claims to have filing cabinets of failed universes. Put in the wrong ingredients, and the universe turns out wrong. For the Millennium Simulation he has got it just right. The secret recipe? Five per cent conventional matter, 22 per cent cold dark matter and 74 per cent dark energy. Quite small variations can completely throw the outcome. Make the dark matter warm, and galaxies don't form. Run side by side on the giant screen, the two versions of the universe at first look pretty much the same, and similar large structures start to form. But soon (meaning 'only' a few hundred million years later) the 'good' universe starts making galaxies and generating the fine structures you see through a telescope; it never happens in the failed one. This seems to confirm that dark matter and dark energy, only quite recently discovered, are key components of the universe. With them in place, the Millennium Simulation has at last shown us how the modern universe evolved.

DATA FILE: DARK MATTER AND DARK ENERGY

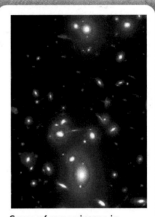

Some of our universe is missing. If you measure how fast a galaxy is spinning, and then work out how much gravity is needed to hold it together, the stars you can see don't have enough mass. The extra gravity is due to stuff we can't see – dark matter – which makes up 22 per cent of the mass of the universe. Recent findings reveal that the expansion of the universe is speeding up. This acceleration is powered by something else we can't identify – dark energy – which makes up 74 per cent of the mass of the universe. Between them, dark matter and dark energy make up 96 per cent of the universe, which means that the matter we can see makes up only 4 per cent of the total. In the image above, dark-matter gravity heavily distorts the light from galaxy Abell 2218.

DATA FILE: PLASMA

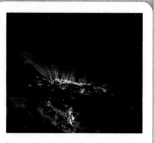

Most matter on Earth is made up of combinations of atoms. But inside stars, at temperatures of millions of degrees, ordinary atomic matter can't exist. Atoms are stripped of their electrons and separate into a sort of electrically charged cloud, leaving the rest of the atom, its nucleus, minus its electrons as a charged 'ion'. Matter becomes a mixture of ions and electrons known as plasma. Plasma is ejected from the Sun (above) when flares erupt on its surface.

'Nuclear fusion is the process that powers the Sun and all the stars. Our aim is to recreate those processes here on Earth'

Andy Kirk
Culham Science Centre

Carlos thinks we have learned more about the universe in the last 10 or 20 years than in the whole of human civilization. He firmly believes in simulated universes. True, they do not have the absolute certainty of a mathematical equation. But what use is an equation about the formation of the universe, unless you can show that it does form a universe? This is also possibly the only way to enjoy the laws of physics in glorious 3D!

CERN is concerned with what happened moments after the Big Bang, and the Cosma computer can take us to the point in universe-formation where galaxies formed. But what about the things in the universe we are familiar with – the stars, the planets and ourselves? How are these made? Stars are the key to all the rest because stars are factories, taking the raw material made in the Big Bang and fashioning from it the chemical elements that make up the rocks and gas of the planets, and eventually the molecules that gave rise to life. Now, building stars is something scientists are doing in the lab.

STAR LABORATORY

At the Culham Science Centre in Oxfordshire they make stars every day – one every 15 minutes, when everything is going to plan. The main ingredient is hydrogen, although star-makers on Earth generally use two heavy forms of hydrogen called deuterium and tritium.

Hydrogen is the simplest and lightest element, and in gas form it was used to lift airships and balloons – until it's extreme flammability proved fatal. Every atom of hydrogen in the *Hindenburg* airship, and all the 'H' in the H_2O that comes out of your tap as water, was made in the Big Bang. But the 'O' in water – oxygen – had to wait until stars first caught fire. On Earth, making stars needs a certain ingredient: hydrogen plasma. We are used to seeing, say, water as either a solid (ice), a liquid (water) or a gas (steam). But there is also a fourth state of matter – plasma. Plasma forms under extreme conditions, such as those that occur inside a bolt of lightning or inside a star (see box).

Magnets can be used to manipulate the electrically charged plasma. The Culham star-making machine squeezes the plasma tightly into a sort of magnetic bottle. It would be more convenient to use a glass or stainless steel one, except that it would be destroyed by the extreme temperatures involved: about 20 million degrees C (36 million degrees F). As the power is cranked up and the plasma squeezed more tightly, it eventually does a remarkable thing: it ignites as a tiny star. The intense glow doesn't last long. For the moment the scientists film it in ultra-slow motion to see how it starts, and more importantly why it breaks down after such a short time.

In the fiery furnace hydrogen succumbs to the heat and pressure, and some of the hydrogen nuclei fuse to make a new element, helium. The energy that is released should, in theory, help to raise the temperature and keep the reaction going – something the Sun and the other stars manage to do for billions of years.

The view on the right is of plasma generated inside the JET nuclear fusion machine at Culham in Oxfordshire. The plasma is held in place by huge magnetic fields. Its temperature reaches millions of degrees. The visible glow is from the outside of the plasma where it interacts with atoms – the plasma itself emits no light.

29

An enhanced-colour image from the Hubble Space Telescope shows the glory of the Crab Nebula – the remains of an exploded star. The purple glow in the middle results from electrons spiralling through the magnetic field surrounding the central pulsar. The nebula is located about 7000 light years from Earth in the constellation Taurus.

At Culham, they are still working on it. But with the fusing of hydrogen into helium, turning one element into another, the process that happens in stars has begun. In the lab, the nuclear fusion goes no further. In stars more and more fusion takes place, making heavier and heavier elements, all the way up to iron.

The Culham group makes stars only as a by-product of their real work – a new clean form of power we can use. If they can make their plasma hot enough, big enough, and last long enough, then the power spewed out by

the fusing hydrogen should be greater than the power needed to run the machine. It is a huge technical challenge, but one that has tantalizing benefits should they succeed. Because, unlike the burning of fossil fuels to make electricity, fusion power does not use up resources we cannot replace, and is clean – the only by-product is helium, which is harmless. And unlike the nuclear power we have now, where atoms are split apart rather than fused together, there is no radioactive waste to store for hundreds of years.

EXPLODING STARS

If stars make the chemical elements, how did those elements end up on Earth? In the days before telescopes, the heavens were at first thought to be pretty much unchanging. A notable exception was the occasional appearance of 'guest stars' as the Chinese astronomers called them. Later, European astronomers referred to them as 'new stars', and that is how they appeared. The most famous guest star was spotted by Chinese astronomers in 1054, and was so bright that it was visible for three weeks *during the day* (see page 130). Over the next two years this guest star faded away, as they all do. Once good telescopes became available, it was possible to work out what had happened, because in the place occupied by the 1054 star you now see something else. Even to the untrained eye, it looks like the aftermath of a huge explosion, and that is what it is. Just as in the lab at Culham, a star burns hydrogen and makes helium. But in a star the temperatures and pressures are far greater, setting up a chain of transformations in which heavier and heavier elements are created, starting with helium and ending in the hot dense centre of the star with iron.

But this is not an endless process. In making the elements, the star is burning up its fuel. The energy released not only makes the star shine, it is what holds the star in shape: without it, and the pressures it creates, the star would collapse under its own gravity. Eventually this is what happens: without enough fuel to burn, the star cannot hold itself up, and from something perhaps many times the size of the Sun (only larger stars can make heavy elements) it collapses. This sounds like an implosion, rather than an explosion, and at first that is true. But a huge shock wave is produced that

DATA FILE: SEEING SUPERNOVAE

Since the supernova explosion seen in 1054 that formed the Crab Nebula, only two have been observed in our own galaxy, in 1572 and in 1604. Both of these were as bright as Venus. More recently, in 1987, a supernova went off in the neighbouring Large Megallanic Cloud and gave astronomers a chance to study the death throes of a massive star.

rebounds outwards, shooting the outer layers of the star into space. There is a tremendous release of energy – what those ancient astronomers saw as a guest star. We now call it a supernova. The conditions are so intense that most of the elements heavier than iron, impossible to produce in a star, are now wrought in the heat and pressure of the supernova explosion. The gold in your jewellery, the lead in weights and old water pipes, the uranium used in nuclear power stations, were all made in exploding stars.

Since we've never delved into a giant element-making star, all this was theoretical until fairly recently. Traces of the elements can be detected in the wonderful colours of supernova clouds, but we couldn't prove what happened in the explosion. Then in 2006 the Spitzer Space Telescope was trained on another wreck of a

The Spitzer Space Telescope is an infrared observatory launched in 2003. It is named after Lyman Spitzer, the first to propose space observatories, in the 1940s.

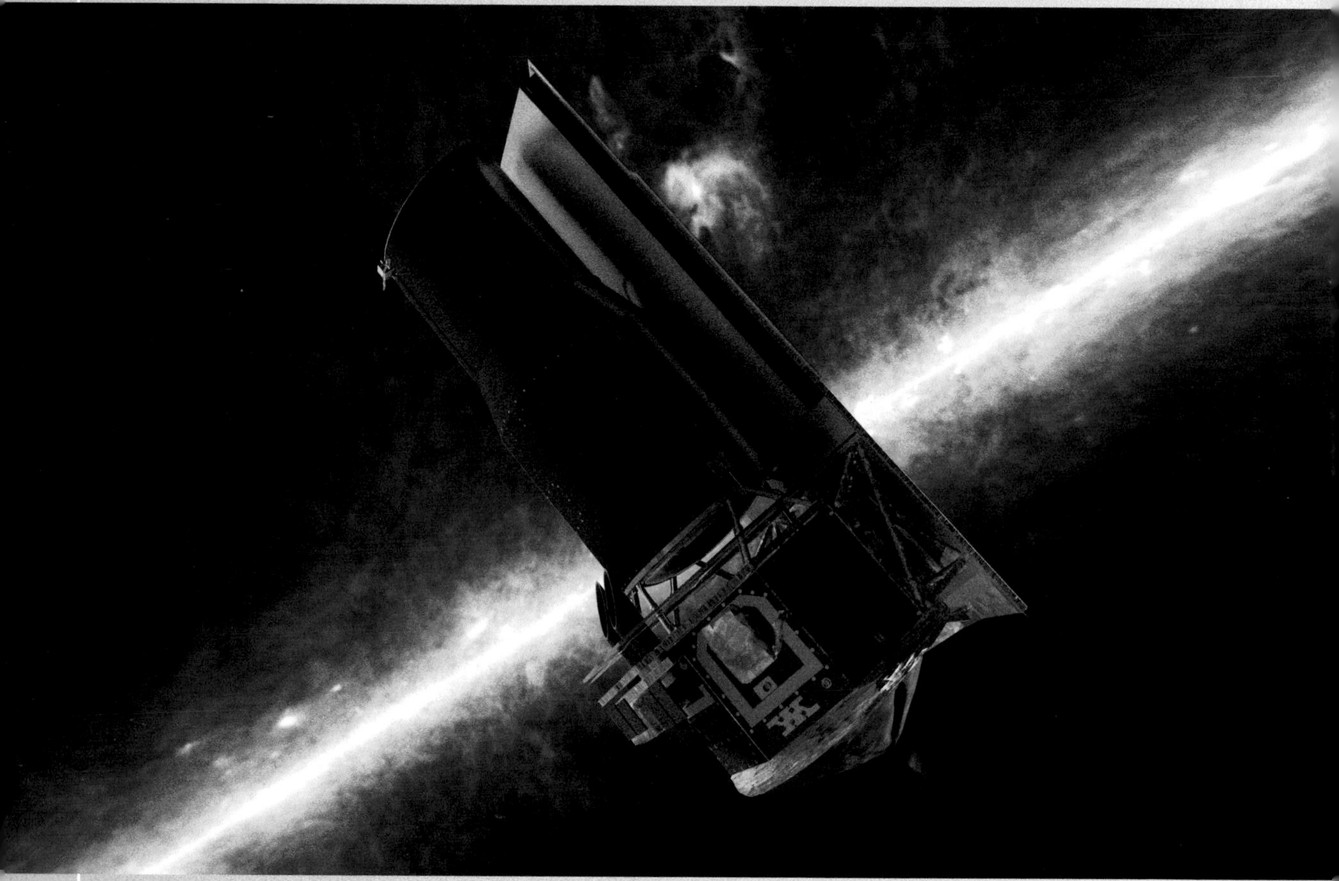

supernova, Cassiopeia A. This one is thought to have exploded in about 1667; it can still be seen expanding, and at 11,000 light years away, within the Milky Way, is fairly close, astronomically speaking. It had been photographed many times before, but Spitzer is an infrared telescope, and can detect cooler matter. When NASA combined the new image with previous ones, it became clear that much of the 'onion skin' arrangement of the original star had been preserved in the explosion, with heavier elements in the centre and lighter ones near the outside. Firework manufacturers tell us that they depend on the same effect – pack layers of different colours around the explosive core, and the order of the layers will be preserved in the final bang.

In Cassiopeia A, the firework display is caused at least in part by the shock wave from the explosion heating up the successive layers in turn, and making them glow with colours determined by the elements they contain. Since there is no prospect within our lifetimes of visiting even the stars of our own galaxy, this colourful cosmic show might have remained the closest we would ever get to the origins of the stuff we're all made of. But although we can't visit stars, what if stardust came to us?

SEARCHING FOR STARDUST

On 15 January 2006, a small capsule re-entered Earth's atmosphere, eventually making a soft landing in Utah, USA. Once recovered, its contents were taken with great care to the cleanest of clean rooms – as it was vital to avoid contamination. This is the Stardust mission, daring but brilliantly simple, and the curious grids extracted from the capsule are thought to contain

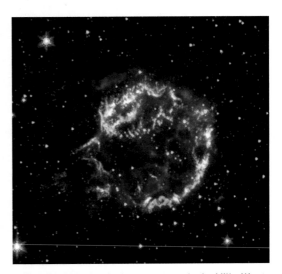

Cassiopeia A was the last supernova in the Milky Way to have been visible with the naked eye. The shock waves are still travelling at millions of kilometres per hour.

the first-ever samples of dust from exploded stars and from comets. They were collected from within the solar system, because it is now realized that although it is rare, interstellar dust should be found almost everywhere. The surprises thrown up by the Stardust mission's comet particles are the latest in a mounting body of evidence that is causing scientists uncertainty about how our solar system ended up the way it has. Some of that evidence comes from close to home, within the system itself. But we are now finding other solar systems – and planets around stars other than our Sun (see chapter 4). The systems found so far are not much like our own. In part this is because the technology used to find them can at present only detect huge planets. But the fact is that these huge planets exist, and it could be that our system with its inner rocky planets (Mercury, Venus, Earth and Mars) and outer giant planets (Jupiter, Saturn, Uranus and Neptune) is the odd one out.

An artist's impression shows the Stardust spacecraft closing in on comet Wild 2. Its dust collectors are catching cometary ejecta for analysis.

STARDUST MISSION

Stardust was launched in 1999 and returned to Earth in January 2006. Its primary mission was to investigate comets, by collecting comet dust as it passed through the tail of comet Wild 2 (pronounced 'Vild 2' because that's how its Swiss discoverer said his name) in January 2004.

Collecting was done with paddles, extended into space a bit like holding a tennis racket out of a car window. One side of the paddle was for comet dust, the other side for stardust. The paddles consisted of metal grids, each square filled with aerogel foam. A type of glass, aerogel is one of the strangest materials made – and the lightest. It nearly floats in air, and is often described as 'solid smoke'. It was chosen because it could slow down and capture tiny particles of comet or stardust without damaging them. The particles travel at several kilometres a second, and would vaporize if they hit anything more dense. Each particle is expected to make a carrot-shaped hole in the foam, with the particle sitting at the point of the carrot. Many comet dust particles have been found, but stardust is much rarer. The task of finding any particles has been described as like looking for 45 ants on a football field.

At the time of writing, we can't say whether the aerogel contains any stardust because the researchers at NASA don't know yet. The particles are so small and so rare that actually finding them within the aerogel foam of the collecting grids will be a mission in itself, and NASA has recruited members of the public to scan high-magnification movies of the gel in the hope of finding one of the estimated 45 tiny particles. If particles haven't been found by the time you read this, and you want to have a go, look at NASA's Stardust@home project on the Internet.

Stardust's main aim wasn't to pick up a few stardust particles, but to collect other material by passing through the tail of a comet. Many comet dust particles have been found and recovered, and are already yielding strange and interesting results. Some of them seem to have been formed at high temperatures, which is odd given the icy nature of comets. Previously it had been assumed that they formed in the cold part of the solar system. Another surprise in the comet particles returned by Stardust is the rich variety of minerals they contain. Comets are fascinating to scientists because they are thought to be fairly untouched relics from the early solar system, bits of leftover material after the planets formed. Studying them might give clues about the materials from which Earth and the other planets were made. The dust collected from Comet Wild 2 contains a variety of organic compounds, molecules made from carbon that also crop up in living things. We don't know for sure how life began on Earth, but it seems that the early solar system had lots of the raw materials that would have made it possible.

The Stardust capsule with its particle samples landed intact in Utah on 15 January 2006. The Stardust spacecraft remains in orbit around the Sun.

A comet particle impact in one of Stardust's aerogel tiles. The impact velocity was 6100 metres (20,000 feet) per second.

This is one of the dust collectors. To analyse the aerogel blocks for dust 60 million photos will be taken.

This digital artwork shows the protoplanetary disc of rock and dust surrounding a new star in an infant solar system. A Jupiter-like planet has already formed close to the star.

'Planets are actually formed in an environment where baby stars are forming'

Helen Fraser
astrochemist

DATA FILE: ZODIACAL LIGHT

Even today, not all the material from the birth of our solar system is soaked up. Some large chunks form asteroids and ice dwarfs, but smaller grains of dust are scattered across the plane of the solar system. Sunlight reflected off this material creates what is known as the zodiacal light, a faint glow along the line of the ecliptic (Earth's path around the Sun). The light can be seen just before dawn or after sunset. Recent observations by a variety of spacecraft have shown structure in the zodiacal light including dust bands associated with debris from particular asteroid families and several cometary trails.

RETHINKING PLANET FORMATION

The origin of planets once seemed straightforward. Young stars had discs of gas and dust around them – material left over from the formation of the star itself. What happened next was down to gravity. As the disc of dust and gas cooled down it condensed – bits clumped together, into particles of rock and metal. The bigger particles became centres of attraction, pulling in smaller ones, and growing all the time. Once the Sun ignited, 5 billion years ago, the part of the disc closer in was warmer and lost its ice. The rocky planets formed there, with the cold giants Jupiter and Saturn forming further out, and Uranus and Neptune, which have both rock and gas, forming further out still. Recently, however, it has been suggested that this might not work. Computer simulations of how gravity could work on the material from which planets are made suggest that rather than a gradual build-up of many lumps, which might then collide and coalesce into planets, it is more likely that planet formation kicks off with a sudden collapse in the

star's gas cloud. This would result in the fairly rapid formation of a really big planet, the size of Jupiter or even bigger. Like Jupiter, this first big planet would be a 'gas giant'. It would have a dramatic effect on the remaining dust and gas, known as a 'protoplanetary disc', sweeping up a lot of material near its orbit. The idea then was that the rocky planets would form from the remaining inner part of the disc, the icy planets from the outer part. But this may not be right either, because there would not be enough material left in the outer part of the disc to make Saturn, Uranus and Neptune.

The new idea is that big planets will form close to their star, and then in the case of our solar system move further out – though it is not clear how this would happen. What we do know is that where planets have been found round other stars, they are very big and very close to their stars, sometimes orbiting in days: so big planets can form very close to a star. This is a distorted picture because with present technology these are the only sorts of planets we can detect – using our technology, aliens in another solar system might not be able to detect any of our planets. There are many more questions to be answered – including how planets like Jupiter could have formed in the hot zone close to the Sun. But as new solar systems are discovered, some of them in the early stages of their formation, the hope is that they will throw light on how our own system was made.

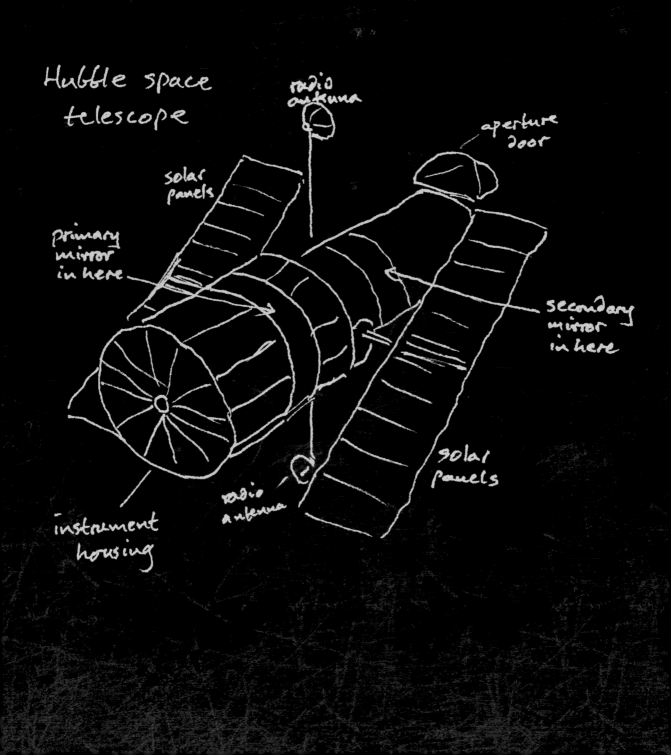

Hubble space
telescope

radio
antenna

aperture
door

solar
panels

primary
mirror
in here

secondary
mirror
in here

solar
panels

radio
antenna

instrument
housing

2/ SEEING THE UNIVERSE

In writing this book we were lucky enough to visit some of the largest scientific machines ever built – many of them located in some truly extraordinary places. But the single most memorable sight of all was perhaps also the simplest and grandest: the night sky.

Just before Christmas 2006, at Paranal in northern Chile, we were about to shoot some night-time telescope sequences for our television series. Leaving the observatory's restaurant, we had to pass through a 'light lock' – double doors that prevent any light escaping. We had been issued with torches because there are no outside lights. This observatory was built on a remote hill in the Atacama Desert because there is so little artificial light anywhere near – light pollution being the enemy of good astronomy. But any thought of using the torches was forgotten the moment the door closed behind us because we were so stunned by what we saw.

We had never seen a sky like it. There was no Moon that night, but the starlight alone gave enough light to see by. There were so many stars that it was sometimes hard to recognize even the constellations we knew. From cities you can usually see only the seven brightest stars of Orion, but here it was displayed in a sea of stars we'd never seen before. The Milky Way, a great bright band of stars, was easy to spot. But most extraordinary were the two misty patches to the right of the Milky Way, the Large and Small Magellanic Clouds. These are named after the great Portuguese explorer Ferdinand Magellan, who saw them on his voyages to the southern hemisphere in the 1520s (though, of course, they must have been seen for thousands of years before him). The 'Clouds' are in fact the nearest galaxies to our own. Seeing them with our own eyes was thrilling.

The experience really impressed us, as it must have impressed millions before us throughout human history. The night sky is so beautiful and profound that you want to find out everything about it. What's out there? What exactly are the dots and patches of light we see? The questions come easily enough, but it isn't clear how you would go about finding the answers. We were fortunate to have an astronomer with us to explain, but what if a person knew nothing about the night sky? How would they even begin to investigate it?

In the dry air over Paranal the bright band of the Milky Way is astonishingly clear and the stars sparkle like coloured jewels. To the right of the photograph can be seen the hazy light of the Large Magellanic Cloud.

This book is possible only because at some point in the past people stopped staring in wonder at the sky and began using their ingenuity to explore it. At first they simply used their eyes, then basic measuring devices. Eventually, about 400 years ago, came the invention of the telescope, which for the first time gave us a view of the universe greater than what our eyes alone could see. The first telescope was crude, but its revelations were profound. Once it was clear that technology could give us new ways of seeing the universe, a race began to make bigger, better and more ingenious devices. In this chapter we bring that story right up to date, visiting some of the machines that are broadening our view of the cosmos way beyond anything our ancestors, staring at the stars, might have imagined.

THE GREAT TELESCOPE RACE

On that starry night in Chile we were visiting a place at the forefront of the latest quest to see further into the universe. Many scientific instruments have mysterious names, but this one is simple: it is the VLT, or Very Large Telescope.

'Very Large Telescope' is more than just a name: it's a statement, too. In astronomy, bigger seems to be better. The push towards larger and larger telescopes began almost as soon as the first simple telescopes became available at the start of the seventeenth century. Curiously, the materials for making a telescope had been to hand since the thirteenth century, when shaped glass lenses for use in spectacles had been invented. It only needed someone to hold up two spectacle lenses of the right sort one in front of the other, and look through them, but this chance discovery seems to

DATA FILE: MEASURING THE COSMOS

The first technology used to explore the universe was very basic, but devices such as the quadrant (above) were important because they allowed early astronomers to record how far an object was above the horizon, and therefore how the stars and planets move from day to month to year. From this followed the discovery that the Sun, not Earth, is at the centre of the solar system, and that the planets move in elliptical orbits. By ingenious use of mathematics and simple Earth-based devices, it was even possible to measure distances in space, including the distance between Earth and the Moon.

have taken nearly 400 years. It is ironic that Hans Lipperhey, the German-born Dutch spectacle-maker widely credited with inventing the telescope, failed to get a patent to protect his invention in 1608 because it was deemed too easy to invent. This may have been because the device for 'seeing faraway things as though nearby' had already been invented several times by craftsmen who made spectacles, but they had been either illiterate or simply uninterested in promoting the invention. Even so, Lipperhey's patent application in The Hague seems to have sparked immediate interest across Europe.

One man very keen to get his hands on a telescope was the Italian mathematician and scientist Galileo Galilei. In the summer of 1609 Galileo was in Venice when he heard that a telescope had been brought to Padua. He rushed there, only to find that the telescope's owner had already gone in the opposite direction, back to Venice. But by then Galileo had gathered enough information to build his own, initially one of three times magnifying power. It worked in the same way as the binocular-like 'opera glasses' you sometimes find in theatres to improve your view of the stage. Their magnification power is similar to that of the first telescope, so that when you use them you have to convince yourself that you are actually seeing any more than with the naked eye. Galileo set about improving the design, and within weeks went public with an instrument that could magnify eight times. Then, in October 1609, he made a telescope of 20 times magnification. This was the instrument that revolutionized astronomy. Almost immediately Galileo turned it towards the Moon, Venus and Jupiter – and made three key discoveries.

This is a modern reconstruction of one of Galileo's telescopes – set against the suitably Renaissance backdrop of the Basilica di Santa Maria del Fiore, Florence.

The early seventeenth century was a time of great upheaval in science. Exciting new investigations and ideas were being fuelled by the rediscovery of the science and mathematics of ancient Greece. The Catholic Church, however, was trying hard to defend the traditional view of the world it had promoted. This was based on the work of the Greek philosopher Aristotle, whose model of a universe made of perfect crystal spheres revolving around Earth was based on philosophy rather than on hard evidence.

Now, for the first time, Galileo had evidence, and he saw that the universe was not as the Church believed it to be. The Moon he saw not as a mysterious, glowing heavenly object, but as a landscape, with mountains and other features that made it clear that this was a rocky world. Venus, which previously had seemed like a bright star, was now revealed to be something else entirely. Galileo's telescope showed that, unlike the stars, Venus was not a point of light but a disc. Further, he noticed that the shape of Venus changes rather as the Moon appears to change through its monthly cycle, going from crescent to half disc to full disc. Jupiter also appeared as a disc, but this time Galileo could see that it was always accompanied by four tiny points of light.

Galileo understood what he was seeing. Venus could change shape in this way only if it was a spherical object going around the Sun, with the shape apparently changing as we see parts of the planet in sunlight or in shade. The little spots of light near Jupiter were its moons. Their changing position from night to night could be explained if these were objects going round the giant planet. Galileo published his findings in the spring of 1610. His observations were a staggering achievement because they showed that the accepted view of the universe was utterly wrong. Heavenly bodies were not all perfect points of light – Venus and Jupiter were clearly lumps of something, as is Earth. Nor is Earth at the centre of the universe

BIOG FILE: GALILEO GALILEI

Galileo was an Italian astronomer, philosopher, mathematician and physicist whose brilliantly original achievements earned him the title 'father of modern science'. Born in Pisa, Italy, in 1564, Galileo was professor of mathematics at Padua University from 1592 until 1610, the year he published the first of his astronomical observations. In 1611 he took his telescope to Rome so that priest-mathematicians of the Collegio Romano could see the moons of Jupiter with their own eyes. By 1633 his publications on astronomy had brought him into conflict with the Church. He was ordered before the Holy Office in Rome and tried by the Inquisition for heresy. Galileo spent the rest of his days under house arrest and died in 1642.

with everything revolving around it on crystal spheres. Venus appeared to orbit the Sun, and the moons of Jupiter (which had not featured in any previous model of the universe) revolved around their own planet, not Earth.

Putting the Sun at the centre of the solar system was not Galileo's idea. That insight had come from the Polish mathematician Nicolaus Copernicus, who found that he could explain the motion of the planets much more easily if he assumed that they moved around the Sun (see page 102). This theory was feared by the Church, so Galileo's use of the telescope to provide hard evidence that Copernicus was right proved very controversial. The world's first decent telescope had been a powerful instrument indeed, and Galileo suffered the consequences – he was tried for heresy in 1633.

Describing Galileo's telescope as having '20 times magnification' is misleading – 20× is the magnification of a pair of powerful modern binoculars, and Galileo's instrument was nothing like as good. The problem was with the design and the materials used. It was probably the improvement of glass lenses that had made Hans Lipperhey's invention of the telescope possible. But the glass was still poor quality by today's standards. The shaping of lenses was also pretty crude, and was really good only near the centre of the lens. So telescopes like Galileo's used an aperture – a disc with a hole – to make the light pass only through the middle of the lens. This, coupled with the arrangement of lenses, gave a very narrow angle of view. Modern attempts to recreate Galileo's telescope result in a device that can see only about a quarter of the full Moon.

This seventeenth-century picture illustrates Copernicus's new heliocentric view of the universe. Earth can be seen moving around a central Sun.

Higher-magnification telescopes of this type were made, but were not much use. They didn't let you see much more, and were incredibly hard to aim at the right point in the sky. Things improved slightly when the first purpose-built astronomical telescopes were built in 1617 and 1645. Telescopes like Galileo's used a positive lens (a magnifying lens) at the front and a negative lens (which makes things appear smaller) near the eye. This arrangement produces an image the right way up. But shortly after Galileo's telescope, this design could be improved no further. It was then found that using two positive lenses instead gave a better result, and in particular a wider angle of view, and this also made it easier to find objects of interest in the sky. The image was upside down of course, but this was a small inconvenience to suffer in return for the better-quality view and much wider angle.

The race to make the world's largest telescope started in earnest in the middle of the seventeenth century. The tubes of Galileo's telescopes were about 2 metres (6 feet) long. By the 1650s, the great Dutch scientist and inventor of the pendulum clock, Christiaan Huygens, had made a telescope that was 7 metres (23 feet) long. Then a third lens was added, and the race continued. Eventually the world's largest was an enormous 42 metres (140 feet) long. After that, it was completely impractical to have a telescope tube at all, so the front lens (the objective) was mounted separately on a tall building.

MAGNIFICATION ISN'T EVERYTHING

The aim of building bigger and bigger telescopes was to see more – either more detail in the objects known about, or to reveal new objects that had previously been invisible. But although magnifications increased to well over 100 times, these telescopes didn't show anything like 100 times more. Even today cheap telescopes are marketed on their magnifying power, and if you buy one you'll be disappointed. Magnification is not everything. If Jupiter appears as a featureless blob in your not-very-good 100-times-magnifying telescope, you haven't gained much. If, on the other hand, you can see swirling bands of cloud and the famous Great Red Spot, then you really are seeing new details. What has improved is the resolving power of the telescope. If my telescope sees one point of light, but yours shows that really we are looking at two stars very close to each other, then yours has the higher resolving power. Resolution was one of the key factors in the progress towards telescopes such as the VLT.

Jupiter and its four largest moons are here seen through a powerful telescope from Earth. The telescope's resolution is sufficient to allow a view of the planet's distinctive honey-coloured bands. Viewed over the course of an hour or so, the moons' orbital movements are clearly discernible. They are called the Galilean moons in Galileo's honour.

Most modern telescopes, including the VLT, do not use lenses at all, for a reason that has its origin in a discovery by the great Isaac Newton. His work with colour led him to understand why the telescope lenses of his day were unsatisfactory. In his most famous optical experiment, Newton showed how white sunlight was split into the colours of the rainbow as it passed through the glass of a prism. He realized that exactly the same thing happens when white light from a star passes through the glass lens of a telescope. The colours in the starlight separate slightly, with the result that it is impossible to focus the different colours at the same point; each comes to a focus at a different distance from the lens. Instead of a point of white, you see a small white circle surrounded by coloured fringes. Although using different glasses and shapes of lens did eventually result in telescopes where the colour distortion (called chromatic aberration) was not such a problem, Newton hit upon a brilliant and entirely different solution.

If the problem happens when light passes through glass, he reasoned, then don't pass it through glass. Newton saw that curved mirrors could do the job just as well as lenses, and invented the reflecting telescope. This was a major breakthrough. If you double the diameter of the mirror in a reflecting telescope, you double the resolution. But as you double the diameter of the mirror, you increase the amount of light it gathers not by two, but by four times. So you can see something that is four times fainter. Its magnificent mirrors help the VLT to see into the distant universe.

THE VLT

The Atacama Desert, home to the VLT, is the driest place on Earth, and it is the inhospitable geography that makes this the ideal spot for a sophisticated telescope. If you are going to look at the sky, you want an uninterrupted view, so height and a place with no brightly lit streets and towns is important. The ultra-clearness of the air here is another factor. This is thanks to the Humboldt Current, a cold stream that runs up the coast of Chile, cooling the Pacific Ocean and keeping clouds trapped down near the coast.

Each evening at sunset, the astronomers at the VLT briefly emerge from their control room to catch the magical atmosphere as the Sun dips below the horizon. The four big telescope enclosures open in turn, pointing to their first targets of the evening. About now the cry 'Earth shadow' goes up, and everyone turns to see the blue curved shadow of Earth cast on the pink sky by the Sun, now just below the horizon. (You can see the Earth shadow in the photograph at the top left of page 52.) It is a good moment to explain why the Very Large Telescope is reckoned to be the best in the world.

Like giant nocturnal flowers unfurling their petals, the enclosure doors of the VLT slide open as the sun sets, ready for the night's stargazing.

DATA FILE: THE VLT MIRRORS

The big mirrors of the VLT are extraordinary. They are 8.2 metres (26.9 feet) across, but only 17 cm (6.7 inches) thick, and are flexible. Rather than being ground from a stiff slab of glass, these bendy mirrors can change shape while inside the telescope. The 45-tonne 'blanks' were cast in Germany from a glass-like material called Zerodur. To keep the mirrors in shape as they solidified, the moulds were spun until the temperature dropped to 800°C (1472°F). The mirrors may have stressed if cooled too fast, so cooling took three months. Then they were carried to an optical factory in France where the surfaces were ground to an accuracy of better than a millionth of a metre. Finally they were taken to Paranal where the aluminium mirror surface was applied in the surfacing factory (above). Over time the reflecting surface degrades so each mirror is removed and resurfaced about once a year.

THE VERY LARGE TELESCOPE

The VLT, run by the European Southern Observatory, has four separate optical telescopes, organized in an L-shaped formation. The four 8.2-metre (26.9-foot) apertures are the largest and best single-piece telescope mirrors anywhere on Earth. Each telescope has a set of instruments for observing at wavelengths from near-ultraviolet to mid-infrared. The VLT uses a range of techniques to observe and photograph objects, including high-resolution spectroscopy and imaging.

It's not surprising, then, that the VLT has made some ground-breaking observations. It scored a huge first in April 2004 by taking the first direct photograph of an 'exoplanet' – a planet outside our solar system. Previous exoplanets had been detected indirectly (see chapter 3). The telescopes are also leading the search to find out what the fate of the universe will be. By looking for distant supernovae – exploding stars at the edge of the universe – to use as reference points, the VLT will be able to measure how fast the universe is expanding.

Sunk into the ground at the VLT is the Biodome, an oasis filled with tropical plants. Here the 150 staff can relax and bathe in the pool.

During the day, the air inside the enclosures is cooled to the predicted temperature at sunset, so that the mirrors suffer no temperature change when exposed to the night air.

Each telescope weighs 400 tonnes but can 'slew' rapidly on its structure to any part of the sky.

Also remarkable about the VLT is the way the telescopes are used – effectively as an astronomy factory. Researchers from all over the world can request observations of particular objects over the internet, and a computer works out which are best suited to the conditions on the night – what astronomers call the 'seeing'. The system means that many astronomers now work from their desks, getting their results the next day.

One of the main reasons for the VLT's success, however, is that it is located in the southern hemisphere, for the simple reason that some of the most important targets can be seen only from down here. A target of particular interest today is the very centre of our galaxy, the Milky Way. The galactic centre is crowded with stars and obscured by gas and dust. But using infrared light the VLT is powerful enough to see through, and has been key in discovering the supermassive black hole at the centre of the galaxy.

The VLT telescopes lock onto guide stars in order to compensate for the motion of the Earth during the night, as seen by the 'star trails' in this picture. Sharing the mountain with the VLT are four small auxiliary telescopes, two of which can be seen here.

TOO MUCH ATMOSPHERE

So far we've given the impression that the bigger you make a telescope, the better it gets. But there is a catch. Every telescope on Earth struggles to see through the atmosphere. The air is full of 'cells' of slightly different temperatures that act like little lenses. The effect is chaotic and makes it impossible accurately to focus light from the stars that passes through, so objects appear distorted. Moving to a desert mountaintop helps, but the bigger the telescope, the more air it has to look through.

For a while, it seemed as though a limit had been reached for telescopes on Earth, so attention switched to telescopes in space. The launch of the Hubble Space Telescope in 1990 opened up a new era in the exploration of the cosmos, where, in the crystal clarity of space, Hubble can outperform the much larger telescopes that have to peer through the atmosphere.

But ground-based telescopes are fighting back – by correcting the distortions caused by the atmosphere. How this was achieved involved some ingenious problem-solving because the distortion is constantly changing as the atmosphere changes. Incredibly, the challenge was overcome with a bendy-mirror system called adaptive optics (see box opposite).

One day adaptive optics will be fitted as standard to all large telescopes. For now it is added as an accessory. You have to use a small auxiliary mirror because it can adapt its shape fast enough to compensate for the rapidly changing distortions of the atmosphere. It would not be possible to move the VLT's 8.2-metre (26.9-foot) mirrors fast enough. The mirror is moved, but much more slowly – only once a second or so, to correct for distortions of its shape caused within the telescope system, such as when the telescope tilts. This is called 'active' optics because the mirror actively changes its shape.

This is one of the VLT's mirror cells seen from behind. The bank of the mirror carries the active optics system that can change its shape.

DATA FILE:
ADAPTIVE OPTICS

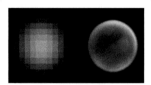

Adaptive optics systems measure the distortion of the atmosphere and then bend a mirror the opposite way in order to correct it. Light from a star should arrive at the telescope in a flat 'wavefront'. Instead, the atmosphere distorts it, resulting in a wavefront like a piece of crumpled paper. Astronomers choose a bright star to act as a 'guide', and analyse its light as it enters the telescope. Wave detectors in front of the telescope mirror work out the shape of the wavefront. This information is fed to pistons behind the correcting mirror that 'adapt' or bend it into a shape that is a 'mirror image' of the wavefront shape. The result is a wavefront free of distortion. To be effective, corrections are made about 1000 times a second. The image top right is of Saturn's moon Titan, taken by the Keck Observatory, Hawaii, using adaptive optics. The image on the left is the same object seen without adaptive optics.

DATA FILE:
VLT AUXILIARIES

The VLT also has four small auxiliary telescopes that can be moved about on tracks. Its greatest trick is to use pairs of telescopes as if they were one big telescope. Two small telescopes 100 metres (330 feet) apart can have the resolving power of a single telescope with a 100-metre (330-foot) mirror, something that cannot currently be built. Light from pairs of telescopes is mixed together in an optical system. What you don't get, however, is a full image – there isn't a full big mirror's-worth of data. The pattern that emerges from the data relates to the size of the object being observed. So for the first time it is possible to measure the size and shape of stars, which in normal telescopes appear as points of light.

Taking pictures of the stars is a technical business. But the sky was so wonderful over the Atacama Desert that Adam couldn't resist trying. This is his picture of the VLT laser, and in it you can also see the Milky Way, and one of the Magellanic Clouds.

On the night we arrived at Paranal, the astronomers were making a star. To use adaptive optics, you need the wavefront of a clear bright star to use as a reference, but it turns out that there aren't enough of them in the sky. Of all possible astronomy targets, only 1 to 2 per cent will have a decent star close enough – one that can be seen at the same time through the telescope. The solution is to make an artificial star, and at the VLT they do it with a bright yellow laser. By a lucky coincidence, about 90 km (55 miles) up in the atmosphere there is a layer of meteorite dust that is rich in sodium. Hit it with a sodium laser, and it glows yellow. The thin yellow beam can just about be seen with the naked eye, but shows up easily in photographs. This gives the telescope a permanently available reference star in exactly the right place.

The Hubble Space Telescope (left) was launched in 1990. Its 2.4-metre (100-inch) mirror may seem fairly puny by today's standards. Yet it is almost the same size as the Hooker Telescope on Mount Wilson in California – the largest telescope in the world until 1948.

HUBBLE TROUBLE

The Very Large Telescope is one of the largest and most sophisticated telescopes ever made. But it is not the one that has captured the popular imagination. That honour goes to a telescope just one-third of the size. When it was completed it wasn't very good, which was embarrassing, having cost billions of dollars. The heroic mission to fix it made for a great story, and probably helped to establish the fame of the Hubble Space Telescope. It is named after the American astronomer Edwin Hubble, who used a telescope almost the same size to find the evidence that the universe is expanding – see chapter 1.

The Hubble Space Telescope is the culmination of a 30-year dream for American astronomers. The building of a space telescope was first recommended by the National Academy of Sciences in 1962; getting it made was a long, slow process. The project got final approval in 1977, the mirror was finished in 1981, and the spacecraft was ready in 1985. The plan had been to launch

the Hubble on the space shuttle in 1986, but the explosion shortly after launch of the *Challenger* shuttle earlier that year put the project on hold. The completed telescope had to wait four years to get into space.

The boast was that the Hubble would have ten times the resolution of any telescope then working on Earth. But once it was in orbit in 1990, NASA discovered the awful truth: undetected since its construction nine years earlier, the mirror had been made slightly the wrong shape, and would be unable to focus images properly. This seems incredible given NASA's reputation for testing everything, but it turned out that the problem was in the very machine used for testing. It was a real blow, a plan took shape to correct the problem. Hubble was to be serviced several times in what was to be a long working life. So the first service mission in 1993 fitted a lens to correct the telescope's vision. Within a very short time Hubble was returning the most stunning images ever seen of the cosmos.

THE HUBBLE SPACE TELESCOPE

Since its launch in 1990, Hubble has proved one of the finest scientific instruments ever made. Its cameras, spectrographs and guidance systems work together to provide dramatic images from the edges of the universe. Several of the HST's systems have now failed, however, and future success will depend upon a service mission in 2008.

1. *Science computer:* controls all the scientific instruments.

2. *Primary mirror:* 2.4 metres (7.8 feet) in diameter, kept at a constant temperature to prevent warping.

3. *Fine Guidance Sensors (FGS):* locks onto guide stars to keep Hubble pointing in the right direction. Important for long exposures such as the 10-day Hubble Deep Field exposure.

4. *Corrective Optics Space Telescope Axial Replacement (COSTAR):* installed in 1993 to correct the defect in the primary mirror.

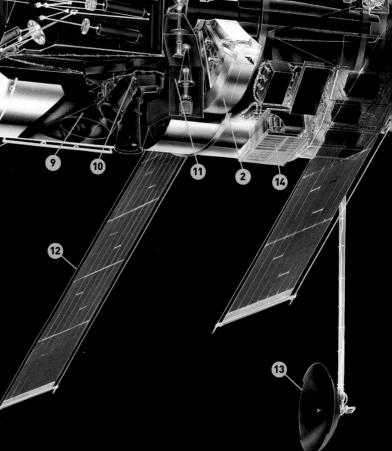

Close up of primary mirror and science instruments.

7. Advanced Camera for Surveys (ACS): three cameras in one, it surveys wide areas of the universe, photographs detail of galaxies, and stars in ultraviolet. Used in the hunt for dark matter, new planets and evolving galaxies. The ACS stopped working in January 2007.

8. Handrails: for use by astronauts on servicing missions.

9. Star tracker: detectors that locate and track bright stars.

10. Gyroscope units: sense changes in the HST's orientation in space.

13. Communications antennas: these provide the link between Earth the HST. Signals are relayed through NASA's Tracking and Data Relay Satellite System.

14. Batteries: six nickel-hydrogen batteries provide power when HST is in Earth shadow. They are recharged with power from the solar panels.

15. Secondary mirror: 0.3 metres (12 inches) in diameter.

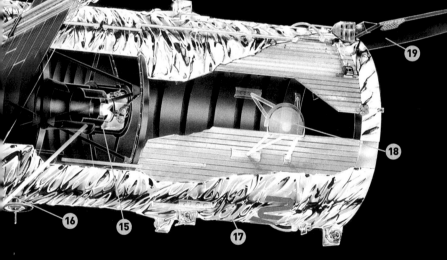

16. Magnetic torquer bars: large bar-shaped electromagnets that react against Earth's magnetic field to help control the Hubble's movements.

17. Insulating blanket: layers of aluminized Kapton, and an outer layer of Teflon help maintain thermal stability.

18. Shuttle support: points of support for the HST when in the Space Shuttle cargo bay.

19. Telescope (aperture) door: closes only when necessary for telescope protection.

5. Space Telescope Imaging Spectrograph (STIS): analyses light by splitting it into its colours. Good for locating black holes. Stopped working in 2004.

6. Near Infrared Camera and Multi-Object Spectrometer (NICMOS): Hubble's 'heat camera' has three different angles of view. Has to be kept cold to avoid detecting heat from the spacecraft. Used for seeing through dust and gas; detecting extremely distant objects red-shifted to infrared.

11. Wide Field and Planetary Camera: known as WFPC2 because the original was replaced in 2002: versatile medium wide camera, has taken many of Hubble's most famous images. Its 48 filters can be used to pick out particular wavelengths, e.g. that of glowing hydrogen.

12. Solar arrays: the main source of power for the HST. They collect solar energy and store it in batteries.

DATA FILE:
HUBBLE THE TEAM PLAYER

Although Hubble can capture ultraviolet and infrared light, it still cannot see all the detail that light can provide.

Other telescopes can see processes that emit X-rays or deeper infrared – but cannot see the visible wavelengths captured by Hubble. So for a complete picture, Hubble images are sometimes combined with those from other space telescopes. In this image of galaxy M82 the visible image is combined with an infrared picture from the Spitzer Space Telescope and an X-ray picture from the Chandra X-ray Observatory. The combined result shows more of what is happening in this galaxy, where star formation happens at 10 times the rate in the Milky Way.

Lars Lindberg Christensen is an image specialist. He turns the raw data into the iconic images that have made Hubble so famous.

HOW TO MAKE A MASTERPIECE

The HST's most famous images have been taken by its Wide Field and Planetary Camera, which only produces black and white images. To turn them into amazing colour, the HST takes three photos through red, green and blue filters in turn. There are more than 40 filters available, so that very specific colours, and wavelengths way beyond normal colour, can be captured. These are then downloaded to Earth, where specialists begin the process of turning the the raw data into wonderful images:

- The black and white images are first adjusted for human vision. In order to capture science data the camera is sensitive to a wider range of tones than the eye can see. These tones are now compressed, so that detail in the image becomes visible.
- Colour filters are then applied. The images are coloured using the same filter colour they were taken through.
- Next, the three images are aligned and stacked on top of one another. This sometimes requires manual adjustment, since the HST's position can move between exposures.
- Finally, the pictures are adjusted until the balance 'looks right'. This is where art comes in. The idea is to show all the important detail – and make a picture that looks stunning.

Natural colour: this picture of the Sombrero Galaxy uses colours chosen to simulate those our eyes would see if we were near enough.

False colour: this helps us to visualize what would otherwise be invisible. Features of Saturn are here seen in infrared light.

This is one of Hubble's most famous images. The 'Pillars of Creation' is a part of the Eagle Nebula where stars are born. Full-size images taken by Hubble's WFPC have a curious 'stepped' shape, because this is the pattern of the electronic sensors: three larger wide-field sensors in an 'L' shape, and the high-resolution Planetary Sensor in the middle.

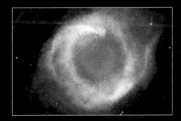

Enhanced colour: by highlighting certain visible colours in an object, subtle details of its composition and structure become apparent.

The Hubble Ultra-Deep Field image is a tantalizing view of thousands of galaxies in the very early universe

Hubble has done so much in the years since the first repair mission that it is hard to sum up what it has achieved. Perhaps its greatest contribution is just being there – reliably turning in the work and being available whenever a new idea needed testing or an event witnessing.

Some of the most beautiful images have come from investigations into the way the universe works, especially in the places where stars are born from great clouds of gas and dust. When the 'Pillars of Creation' (see page 56) picture was first released in 1995, it was front-page news. What was it about this picture of the Eagle Nebula that so captured the imagination? For scientists, Hubble was showing in action something that had been speculated about for a long time: young stars emerging from the surrounding cloud by a process they called 'photoevaporation', where light from nearby stars boils away the gas. The 'Pillars of Creation' showed us a dramatic, dynamic universe far removed from what we can see by staring at the sky even on the clearest night. Hubble has truly given us new eyes for looking at the universe.

Perhaps the Hubble image that has most impressed both of us is less immediately spectacular, but has breathtaking implications. For ten days between 18 and 28 December 1995 Hubble stared at what was apparently empty space, taking a series of exposures of between 15 and 40 minutes. A very narrow view was chosen, selected to have hardly any nearby stars in it. The resulting pictures showed that this patch of deep space was far from empty. In every part of Hubble's narrow view there was a galaxy. Some of them are classic spirals and ellipses, while others are shapes never before seen. There are about 1500 galaxies in the picture, and each could contain 100 billion stars. There is every chance that the same would be found in every other part of the sky. It is perhaps the picture that best shows how truly vast the universe is. Recently, Hubble has done it again, snapping the 'Ultra-Deep Field' (HUDF) in 2004 using the newly installed Advanced Camera for Surveys. The 10,000 galaxies of the HUDF include the youngest galaxies ever seen, born perhaps 400 million years after the Big Bang – the birth of the universe. The light that reached Hubble during the million-second HUDF exposure started its journey 9000 million years before Earth was formed.

THE SECRETS OF LIGHT

Back in 1608, Hans Lipperhey's patent application for his telescope described it as a device for 'seeing faraway things as though nearby'. For the first telescopes this meant seeing things too faint for the naked eye. But the desire not simply to observe the universe, but to understand how it works, has driven the discovery of amazing new ways of seeing that are nothing like the way we see faraway things. These discoveries began when Isaac Newton conducted that experiment to show how white sunlight is actually a mixture of all the colours of the rainbow, and that these can be seen separated out when passed through a glass prism. The experiment revealed something astonishing about the nature of light. You could also think of it as a revelation about the Sun.

Newton could not have predicted that two of his great achievements, the reflecting telescope and the discovery that a prism splits light into its colours, would one day be combined to reveal not only what is happening inside the Sun, but in planets, clouds of gas and distant stars and galaxies. Eventually, telescope and prism would even reveal that the universe is expanding (see page 16).

Revealing as the Sun's spectrum was, some of its secrets remained hidden until optical technology allowed it to be spread out even further. Then something odd appeared: black lines. First noticed by the English chemist William Hyde Wollaston in 1802, these black lines were studied properly by the German scientist Joseph von Fraunhofer who, unaware of Wollaston's work, rediscovered them 12 years later. He didn't know what they were, but this didn't stop him mapping them, finding 570 in all. The letters A to K were the designations he gave to the more distinct ones, sometimes labelling close pairs with numbers, such as D1 and D2. But what were the black bands in the rainbow, and what did they tell us about the Sun?

'In the beginning of 1666, I procured me a triangular glass-prisme, to try therewith the celebrated phenomena of colours'

Isaac Newton

Falling rain produces exactly the same effect as a glass prism, splitting sunlight into its constituent 'rainbow' colours.

DATA FILE:
FRAUNHOFER LINES

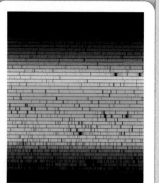

This is a high-resolution spectrum of the Sun. It was made by spreading the light into a 'rainbow', running from top to bottom of the picture, then spreading it out again with another prism from side to side. The 'red' part of the original rainbow is spread into all the red colours from left to right. The dark Fraunhofer lines are now known to be absorption lines where a chemical element, either in the outer atmosphere of the Sun or in the Earth's atmosphere, absorbs light at a very precise and characteristic wavelength. Fraunhofer lines like this may reveal chemicals in the atmospheres of planets orbiting distant suns, as the light passes through those planets' atmospheres on its way to Earth.

The German scientist who worked out what was going on has one of the most famous names in chemistry: Robert Bunsen, the man who perfected the gas burner used in school laboratories. Bunsen used to heat things in a burner, and then look at the glow with a prism, though he wasn't the first to do this. It turned out that if you stick, say, a bit of sodium in the flame, you get a distinctive orangey yellow glow. You will recognize the sodium colour when salty water meets a gas flame as your spaghetti boils over; it is also the colour of sodium streetlights. Seen through a prism, the sodium light is in two distinct yellow lines, very close to each other. Bunsen stuck other elements into the flame, and each produced one or more coloured lines. The pattern is so distinctive and precise that this 'spectral fingerprint' can be used to identify all kinds of mystery substances.

Two extraordinary conclusions immediately followed. If you examine the spectrum of sunlight in great detail, it turns out not to be a smooth blend from colour to colour. Spread it out enough, and eventually it separates into lines. Two of the brightest, D1 and D2, are yellow lines at exactly the same position as the lines produced by burning sodium in a flame. Conclusion one: the yellow lines in sunlight are there because sodium is being burned in the Sun, and indeed the entire spectrum can be accounted for by chemicals heated in the Sun. This was chemistry done at a distance of 150 million km (93 million miles). Then Bunsen compared the coloured lines from glowing chemicals with the black Fraunhofer lines. If he burned iron in the flame, it produced a line where the black E2 line was. Sodium matched the D lines, calcium matched the H and K lines. Conclusion two: The black lines are caused by absorption of light by chemical elements in the Sun's atmosphere. Each element has a specific set of absorption lines, which therefore act as fingerprints for identifying the element.

Once astronomers understood how to use light to fathom the chemistry and physics of the Sun, they could do the same thing for everything else seen with a telescope. At first it was little more than sticking a prism onto the eyepiece, but eventually spectrographs enabled the light from a single star to be analysed. Today every major observatory, including the Hubble and the VLT, has a spectrograph.

Using a spectrum to do chemistry at a distance of billions of kilometres was only the beginning. It turned out to be the key to measuring movement in the universe and to show that the universe is expanding (see chapter 1). More recently, the spectrum has been used to discover planets around distant stars, today one of the really hot areas of astronomical research (see chapter 4).

INVISIBLE RAYS

At about the time William Hyde Wollaston noticed the black lines, the spectrum was revealing yet more secrets to the great German-born British astronomer William Herschel. Herschel was puzzled by the question of the Sun's heat. Where did it come from? Which, if any, of the colours of the rainbow was hot? He set up a prism in bright sunlight, and, placing his thermometer on the spectrum of colours, Herschel measured the temperatures of each. Then he took a break, and on returning found that because the Sun had moved in the sky, the thermometer was no longer in the spectrum; it was somewhere off the red end of the rainbow. But something odd was happening: the thermometer now registered a much higher temperature than when it had been in the colours. He concluded that the heat from the Sun was contained in a sort of light, but one invisible to our eyes. He coined the term 'calorific rays', but from the late nineteenth century the name infrared has been used.

Just a year after Herschel discovered his heat rays, another accidental discovery was made at the other end of the spectrum, when it was found that rays beyond the violet end of the rainbow were particularly good at darkening light-sensitive paper. These were ultraviolet rays. Both infrared and

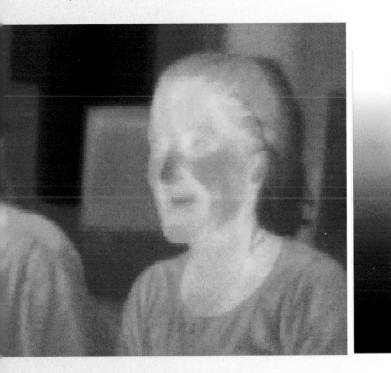

°C

42
40
38
36
34
32
30
28
26
24
22

This image is taken in mid infrared ('thermal') light. Infrared radiation is of a wavelength longer than visible light. The name means 'below red', red being the colour of visible light at the longest wavelength. Astronomy in infrared light can penetrate dusty regions of space such as nebulae.

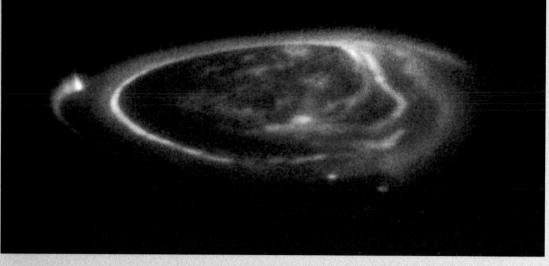

This is the aurora at Jupiter's north pole seen in ultraviolet light by the Hubble Space Telescope. Ultraviolet is radiation with a wavelength shorter than that of visible light. The name means 'beyond violet', violet being the colour of the shortest wavelength of visible light.

DATA FILE: ELECTROMAGNETIC WAVES

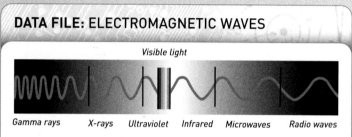

Visible light

Gamma rays   X-rays   Ultraviolet   Infrared   Microwaves   Radio waves

Astronomical objects can emit radiation across the whole electro-magnetic spectrum, from radio waves, through visible light to gamma rays. Some objects shine at each of these wavelengths. Very hot or violent sources emit high-energy gamma rays and X-rays. Cool objects are visible at longer wavelengths. To detect this radiation a range of instruments is used. Some wavelengths do not penetrate Earth's atmosphere and can be detected only by space telescopes.

eventually ultraviolet opened up new areas of astronomy, transforming our views of planets, stars and galaxies. Astronomy is no longer dictated by the sensitivity of the human eye – which evolved for use in sunlight under a blue sky.

When the Space Shuttle visits Hubble for the last time in 2008, one of the crew's tasks will be to fit a new camera more sensitive in both ultraviolet and infrared light. The new James Webb Space Telescope that takes over from Hubble in about 2013 will operate mainly in the infrared. It is still a thrilling experience to take a telescope out on a

dark night and to see through it a mind-blowing number of objects invisible to our eyes alone. But it is worth remembering that even with a telescope our eyes can reveal only a fraction of what the universe contains.

THE FIRST LIGHT

Taking shape in a factory just behind the beach at Cannes in the south of France is one of the most sensitive cameras in the world. It will attempt to photograph the most distant thing *it is possible to see*. The European Planck mission is the third spacecraft to attempt to record the oldest light in the universe. The team behind the much cruder first attempt, by the COBE spacecraft in the 1990s, received the 2006 Nobel Prize for Physics. This light is the holy grail for astronomers because its fine details, which Planck hopes to pick up for the first time, will reveal clues about how we got from a tiny, hot universe just after the Big Bang, to the one full of galaxies that appeared some time later.

When you look in a mirror 5 metres (16 feet) away, you see yourself 3 nanoseconds ago. When you look at the Sun 150 million kilometres (93 million miles) away, you see it as it was 8 minutes ago. Because light takes time to reach our eyes, the further away we look, the further back in time we look. Ironically this means that as we look out towards the edge of the universe, we are seeing the universe not at its biggest, but at its smallest. Light from the very edge is light that started its journey soon after the Big Bang. It would be nice to think that one day we could see back to the Big Bang itself, but it seems that is impossible for the simple reason that in the very early universe there was no light.

The most accepted idea of the very early universe is something called the 'Hot Big Bang' (see page 19). Immediately after the bang, the universe started inflating like a balloon. It consisted of a very hot soup of particles: electrons, found today around the outside of atoms and flowing in electric circuits; protons, found in the middle of atoms; and photons, the particles of light. But in the early years, these particles all interacted together in a way that made the universe opaque (light couldn't get through) and trapped the photons, so there was no 'free' light. But as the universe expanded, it cooled down. Fridges use expanding gas to keep your food cold; expanding universes also get cold. When the temperature cooled to 3000 Kelvin (3000 degrees above absolute zero) the universe lit up: the electrons and protons got together to form hydrogen, and the photons were free to become visible light. This happened when the universe was just 380,000 years old. Because space is still expanding today, the light in it is also expanding and getting longer and longer in wavelength. That first, ancient light should now be visible as microwaves, known as the Cosmic Microwave Background.

Planck (above) is the most sensitive microwave camera ever built. Its detectors have to be kept extremely cold, just a tenth of a degree above absolute zero, and colder than deep space. If it were any warmer, it would radiate heat and be able to see only itself. Most of the Planck spacecraft is a series of fridges within fridges. The telescope is made up of two mirrors that catch microwaves and focus them onto the detectors.

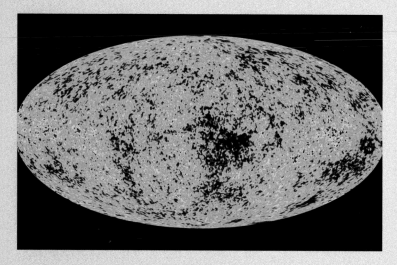

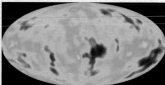

These are the ripples from the earliest light of the universe, seen by the COBE spacecraft in 1995 (top) and at higher resolution by WMAP in 2003 (left).

Space has continued to cool ever since, and the background temperature has reduced from 3000 K to 3 K – just 3 degrees above absolute zero. This first light of the universe was seen by accident in the 1960s, and seems to come from all over the sky, as you would expect. Its discovery was incredibly important since this was the first hard evidence of the Big Bang, which until then had remained a theory. But soon it was realized that this Cosmic Microwave Background (CMB) might actually contain some information about how the universe formed. The question is: what makes galaxies? At some point matter in the universe started clumping together, and galaxies formed, but it is not clear exactly how or when this started.

This is where the Planck spacecraft comes in. Like its predecessors, COBE (for Cosmic Background Explorer) and WMAP (Wilkinson Microwave Anisotropy Probe), Planck will look for tiny variations in the CMB. COBE and WMAP both detected 'ripples in the universe' – hints that even at 380,000 years old it was not entirely smooth – showing that there was some structure. Planck will try to get a more detailed map to see if the ripples can give clues about how the universe we see first emerged from the particle soup. We haven't nailed down exactly how far this means we can see. The light Planck is picking up was released 13.3 billion years ago. If the universe were stationary, then you could say the light travelled 13.7 billion light years. But the universe has been expanding all the while, which puts the distance at about 46 billion light years.

All these ways of seeing the universe depend on things that shine. It might be in radio waves, infrared, ultra violet or X-rays, but to see it, the thing has to shine. However, in the last few years it has been found that most of the universe – about 95 per cent of it – does not shine. This mysterious phenomenon has been given the names dark matter and dark energy; neither can be seen nor even detected (see page 27). Black holes trap light and are also invisible. But resourceful scientists are presently at work on some enormous detectors that may be able to detect even some of the dark stuff, because they see gravity.

THE SOUND OF THE UNIVERSE

Although Isaac Newton's famous work told us what gravity does, he did not explain what it is. Particularly bothering was the implication that gravity acted instantly: if you could somehow conjure a second Sun in the sky, then in Newton's world Earth would feel its pull immediately. Albert Einstein didn't like the idea of something acting instantly at a distance, and instead thought of gravity as acting locally on space. He imagined massive objects making a 'dent' in space into which others could fall. Although the 'dent' would be deepest close to the object, it would reach across the universe so that Earth is attracted even by distant stars. If a new large object were created, or an existing one moved, Einstein predicted, it would create ripples across the universe, travelling at the speed of light. Unfortunately, he also predicted that by the time they reached us the ripples – called gravitational waves – would be far too faint to detect. Rather than a prediction of failure, some people have taken this as a challenge.

The site of the European Gravitational Observatory (EGO) near Pisa in Italy is a real contrast to those selected for optical telescopes. It is on a flat plain, rather than a mountain; it's in a frequently cloudy area, and close to a town. In fact the main requirement was to find a patch of flat ground measuring 3 km (1.8 miles) square, because the Virgo detector is huge. The principle of detecting gravitational waves is very simple. As they pass through Earth, the waves slightly change the length of everything. Imagine a football pitch. As a gravitational wave goes through it is stretched first one way, and gets shorter and fatter; then the other, and gets longer and thinner, before returning to its original shape. So gravitational-wave detectors have two arms at right angles, with sensitive equipment measuring the length of the arms. If this sounds simple, it isn't. Einstein was almost right, because the expected change in length of each of the 3-km (1.8-mile) arms will be the same as one-thousandth the diameter of a hydrogen atom nucleus; in other words, almost too small to measure.

'It is like having seen the universe only with your eyes, and suddenly being able to hear it'

Dr Giovanni Losurdo

The arms of Virgo's gravitational-wave detectors are housed inside two long tunnels on a flat plain near Pisa, Italy.

65

GRAVITATIONAL WAVES

To detect gravitational waves Virgo uses lasers reflecting off mirrors at the end of each arm. By mixing together the signal from the two arms it can detect any change in length. The problem is noise – vibrations that are far easier to detect than a gravitational wave. For now, the team is working to improve sensitivity, and exclude noise.

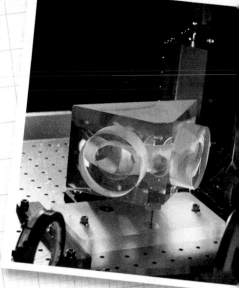

The mirrors (photo, above) are suspended on a system of wires, the lasers are ultra stable and the optics are in cooled vacuum tubes.

VIRGO
Gravity-wave detector

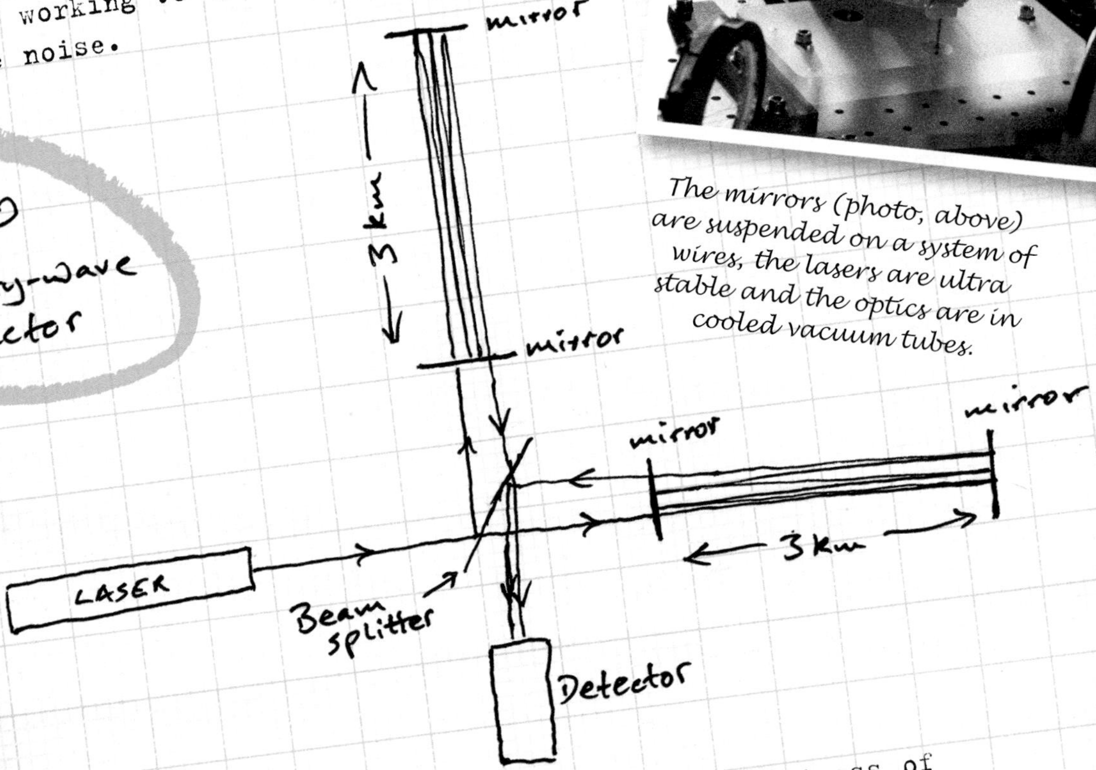

mirror

3 km

mirror

mirror

mirror

3 km

LASER

Beam Splitter

Detector

By 2009 the team at the EGO hope to be in the business of detecting gravitational waves for real. If they succeed it will be revolutionary. The detectors will give access to events and times not available to light astronomers, such as what happens inside a black hole, or in the depths of stars. Gravitational waves may even give an insight into the so-called 'dark ages' of the universe – the time before the first galaxies and stars lit up. As Professor Kip Thorne of the American LIGO detectors says: 'It'll be wonderful ... we'll be seeing the universe and the fundamental laws of space-time in a manner that we have never seen before.'

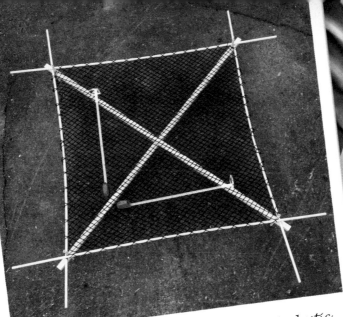

In our gravity wave detector, the black elastic netting is the surface of Earth, with two measuring devices - in this case tape measures - mounted on it at right angles.

Curiously, you will actually be able to hear some gravitational waves. Just as radio waves carry speech, so gravity waves will carry the signatures of events such as the colliding of two neutron stars, and some of these events happen at sound frequencies.

As a gravitational wave comes through, it distorts Earth and everything on it. First it squeezes. then it stretches. As this happens, the tape measures change length.

In our detector, the arms changed length by a few centimetres. In the Virgo arms, the change will be equivalent to one thousandth the diameter of a hydrogen atom nucleus!

67

Mars Science Laboratory

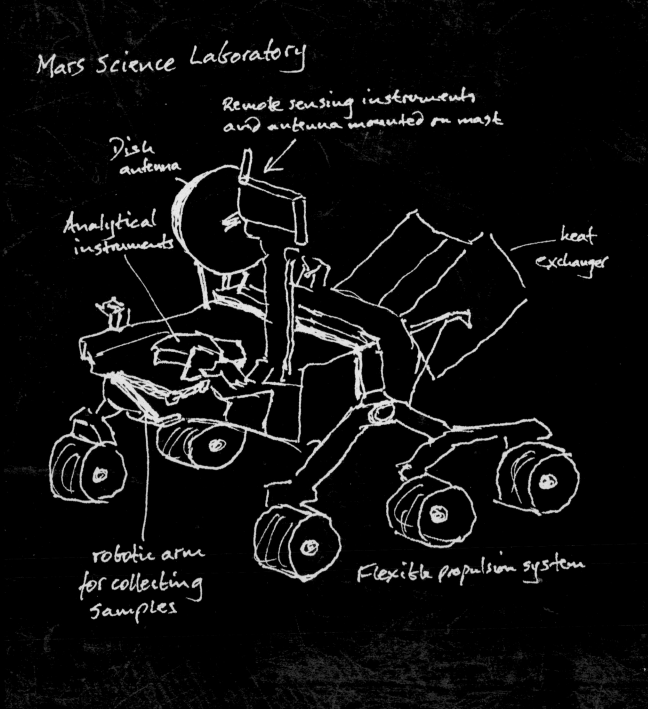

Remote sensing instruments
and antenna mounted on mast

Dish
antenna

Analytical
instruments

heat
exchanger

robotic arm
for collecting
samples

Flexible propulsion system

 3/ SPACE EXPLORATION

'I believe that this nation should commit itself to achieving the goal, before this decade is out, of landing a man on the Moon and returning him safely to Earth. No single space project in this period will be more impressive to mankind, or more important for the long-range exploration of space; and none will be so difficult or expensive to accomplish.' Thus spoke John F Kennedy, thirty-fifth president of the United States, in his State of the Union speech to Congress on 25 May 1961. This was a bold, dramatic and, no doubt to some people, foolish announcement – but it came to pass, and in 1969 American astronauts did indeed walk on the Moon and return safely to Earth. On 20 July Neil Armstrong stepped onto the lunar surface and said, in a muffled voice, 'That's one small step for a man, one giant leap for mankind.'

Kennedy was right: this was the most exciting event in the whole history of space flight. There have been hundreds of other missions, but nothing to make people sit up all night, glued to their television screens. Adam was in Canada at the time, and still remembers those grainy black-and-white images of Buzz Aldrin leaping about with glee.

BIOG FILE: ROBERT GODDARD

Robert Hutchings Goddard was born in 1882, and began experimenting with rockets at Clark University in about 1910. He had great difficulty getting money for his research – no one could understand why rockets might be interesting or useful, but luckily the Smithsonian Institution came to the rescue and supported him financially for many years. They also published his classic 1919 paper 'A Method of Reaching Extreme Altitudes'. On 16 March 1916 Goddard launched the world's first liquid-fuelled rocket after patenting the idea two years earlier. He eventually took out more than 200 patents, but when he mentioned the possibility of sending a rocket to the Moon he was ridiculed by the press. Goddard died in 1945.

To get to the Moon the Americans needed powerful rockets, and fortunately they had a head start. The Chinese invented rockets many hundreds of years ago, but the pioneer of modern rocket science was an American from Worcester, Massachusetts – Robert Goddard.

Goddard was remarkable in that he not only battled on without much support and got his rockets to work, but also pioneered and demonstrated a range of new technologies. He developed pumps to move the liquid fuel, stabilizers and steering mechanisms, and other innovations.

He showed mathematically that a rocket engine would work in the vacuum of space and would not need air to push against. This is what Newton's third law of motion is all about – action and reaction are equal and opposite – but it is nevertheless counter-intuitive. Goddard went on to show that it works in practice.

The British Interplanetary Society was formed in 1933 by a Mr P E Cleator, and remains the world's oldest society devoted entirely to promoting astronautics and space travel. In the late 1930s it published a detailed plan for a manned mission to the Moon, and in 1978 a feasibility study for an interstellar mission. It also publishes the magazine *Spaceflight*.

Meanwhile the German rocket society, the Verein für Raumschiffahrt, was formed in 1927, and the German Army started a rocket programme in 1931. The leading German pioneer was Hermann Oberth, who in 1930 was assisted in rocket-fuel tests by an 18-year-old student called Wernher von Braun.

As the world war loomed in the late 1930s all civilian rocket work was stopped, and Wernher von Braun became technical director at the army rocket centre at Peenemünde on the Baltic coast. Here, using Goddard's published material, von Braun developed Hitler's second vengeance weapon, the *Vergeltungswaffe-2*, or V-2 for short. (The first vengeance weapon, V-1, had been the notorious 'doodlebug' bomb, the first guided missile used in war.)

The V-2 was a remarkable achievement. Standing 15 metres (50 feet) tall, it weighed 12 tonnes, achieved speeds of 5600 kph (3500 mph) and reliably delivered a tonne of high explosive to a target 800 km (500 miles) away. It was used in anger for some six months from September 1944. The V-2 rockets were built at a slave-labour factory called Mittelwerk by inmates of concentration camps, and the appalling fact is that although the rockets killed 7000 people in explosions in London, more people died in Germany during their construction.

After the war the Americans retrieved many V-2 rockets from Germany and launched them in tests at White Sands, New Mexico (seen here). The Russians also captured a number of V-2s.

BIOG FILE: WERNER VON BRAUN

Wernher Magnus Maximilian Freiherr von Braun was born in Wirsitz, now in Poland, in 1912. His father was a senior politician in the German Weimar Republic, while his mother was related to European royalty. She gave him a telescope, and he developed a passion for space, which was further fuelled by reading the fictional works of H G Wells and Jules Verne, and a serious book about rocket science by Hermann Oberth. At the age of 12 he got into trouble with the police for fixing fireworks to a toy car. During the Second World War he developed a guided missile rocket, the V-2, for the Nazis, surrendered to the Americans in 1945, and became the architect of NASA's space programme. He retired in 1972 and died in 1977.

Verne's From the Earth to the Moon *is one of the first stories of the science fiction genre. Interestingly, the story contains many similarities with NASA's Apollo missions, including blasting off from Florida and splashing down in the ocean.*

'We knew that the Russians were going to do it! We have the hardware on the shelf. For God's sake, turn us loose and let us do something. We can put a satellite up in sixty days ... Just give us the green light!'

Wernher von Braun

When the Soviet army was only 160 km (100 miles) from Peenemünde in 1945, von Braun made sure that he and many of his staff were able to travel to the American front, and surrendered themselves there. As a result, the Americans were able to take much of the knowledge and the skills that von Braun had developed, plus a number of V-2 rockets and parts.

The Americans should therefore have had a commanding lead in rocket science, but the government was not interested, and failed to press home their advantage. Even though von Braun was able to build a few rockets, there was little encouragement from the top brass until they received a terrible shock, on 4 October 1957. This was the date the Russians launched the first artificial satellite, Sputnik 1. Adam was at school at the time, and remembers having to write a poem about Sputnik.

The American government was horrified. If the Russians could put a satellite into orbit they clearly had much better rockets, and could therefore rain down nuclear weapons on American cities, while the Americans would not be able to retaliate. And then there was the dreadful publicity – Russia wins space race; United States left behind.

Stung into action, the American government created NASA – the National Aeronautics and Space Administration – on 29 July 1958, and when he came to office, the thrusting young President Kennedy knew he had to get things moving fast. That was why he made the dramatic commitment to go to the Moon.

FICTION

Spaceflight has long been a staple of science fiction, starting with magazine stories and then books in the nineteenth century: Jules Verne wrote *From the Earth to the Moon* in 1865, H G Wells wrote *The War of the Worlds* in 1898; and Hugo Gernsback founded *Amazing Stories*, the first magazine devoted to science fiction, in 1926. More recently, Arthur C Clarke, Douglas Adams, Robert Heinlein and many others have kept those spacecraft zooming across the universe.

There have been radio series, such as *Journey into Space*, which Adam remembers listening to at school; television series, notably *Star Trek*; and films, from *2001: A Space Odyssey* to the six-film *Star Wars* sequence, which in turn has spawned more films, television series, games, comics and so on.

The immense popularity of spaceflight in fiction shows that the idea is profoundly appealing to humans. Some of them at least long to escape from Earth and travel through the cosmos – and a select few have been able to do just that.

FACT

The first person in space was the Russian cosmonaut Yuri Gagarin, who flew once around the Earth in April 1961, became a world hero, and underlined the Russian lead in the space race. The first manned American flight, on 5 May 1961, was not so spectacular. Alan Shepard was the 'pilot', although since he was strapped in place with a steel plate in front of his face he had no control over the spacecraft. It was only a 15-minute sub-orbital flight – orbital flights came later – and as a result of several hours of delays during the countdown, he became desperate

Yuri Gagarin is strapped in and ready for the countdown in the cabin of Vostok 1, the spacecraft that took him into orbit on 12 April 1961.

The Command Module of Apollo 13 is lifted safely onto the deck of USS Iwo Jima after the mission's brush with disaster. The astronauts were lucky that the explosion happened on the first leg of the journey when they had the maximum amount of supplies and power to use in an emergency.

to pee. Ground control were worried that if he wet his space suit he might cause an electrical short between all the electrodes attached to his skin, but on the other hand to go through the lengthy process of extracting him from the cabin in full view of the world's press was not appealing. 'Go in the suit,' they said. He did, and since he was lying on his back in a foetal position, the urine trickled round his waist, and he was launched into space lying in a pool of his own pee.

After a start like that, things could only improve. The Mercury missions with a single astronaut were succeeded by Gemini missions with two, and then – after Kennedy's proclamation – by the Apollo missions, whose goal was to explore the Moon. Apollo 8 orbited the Moon, and Apollo 11 took Armstrong and Aldrin all the way.

There were a few disasters. In January 1967 there was a fire in an Apollo capsule on the ground and

three astronauts died. In April 1970, Apollo 13 suffered a catastrophic leak in an oxygen tank caused an explosion when it was halfway to the Moon. Using extraordinary ingenuity, mission control managed to get the crew safely home after a single loop around the Moon.

By and large, however, the Apollo missions were astonishingly successful. They put the Americans back in the lead, but more importantly, they pushed technology far beyond what had been its limits; the missions showed what could be done, given the will and the money.

The Apollo missions were made possible by means of the vast Saturn V rocket, which was as high as London's St Paul's Cathedral. This was the largest and most powerful rocket ever built, and was the culmination of the work of Wernher von Braun, who thereby achieved his life-long ambition to contribute to space exploration.

The Saturn V rocket launched the first manned lunar landing mission, Apollo 11, on 16 July 1969. Apollo 11 had three parts: its Command Module, the cone at the top of the rocket below the escape tower; the Service Module, the grey cylinder below that, and the Lunar Module inside the white flared faring below the Service Module.

DATA FILE :
SATURN V ROCKET

Launched: 13 times between 1967 and 1973; no failures.
Height: 111 metres (365 feet)
Diameter: 10 metres (33 feet)
Weight: 3000 tonnes; payload (to Earth orbit) 100 tonnes.
Three stages:
First stage: 34 MN thrust, burn time 150 sec, reaches 60 km (37 mile) altitude and 8500 kph 5300 mph;
Second stage: 5 MN thrust, burn time 6 minutes, reaches 185 km (115 miles), 25,000 kph (15,500 mph);
Third stage: 1MN thrust, burn time 2 min to Earth orbit, 5 min to escape velocity.
Fuel: petrol and liquid oxygen for stage one, liquid hydrogen and liquid oxygen for stages two and three.

ESCAPE VELOCITY

The idea of having multi-stage rockets was originally developed by Goddard. In order to escape from Earth's gravity, a rocket has to reach a speed of over 40,000 kph (25,000 mph), or 10 km (6 miles) per second, which requires a huge acceleration and therefore an enormously powerful engine. Calculations show that even with the most efficient engine and the most powerful fuel available, it is not possible to lift a single-stage rocket and all the necessary fuel with enough acceleration to achieve this escape velocity.

The problem is that when most of the fuel has been used, the huge empty fuel tank is entirely useless; it merely adds weight and slows everything down. What is needed is a vast first stage that will lift everything perhaps 65 km (40 miles) and achieve a velocity of perhaps 8000 kph (5000 mph) before all its fuel is used up. The first stage then separates, and the second stage, with much less weight, starts at 8000 kph (5000 mph). This takes the lighter vehicle to 200 km (120 miles) and 25000 kph (15,000 mph) before separating. The third stage is then able to achieve escape velocity.

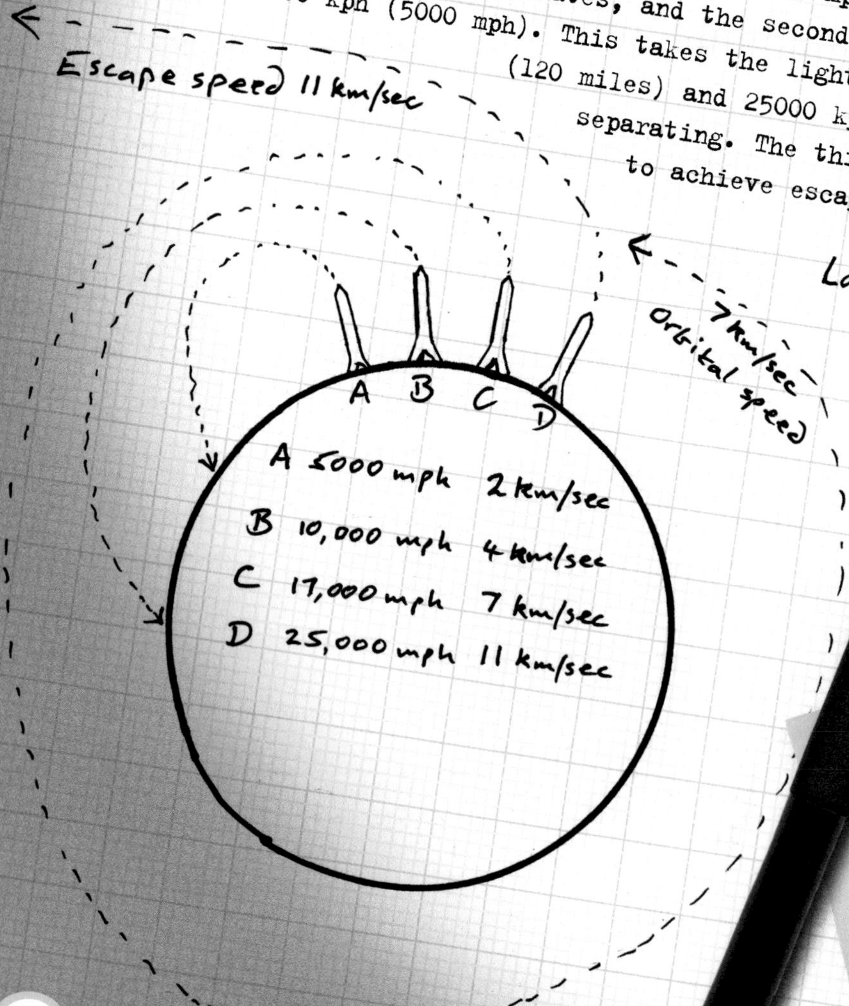

Escape speed 11 km/sec

7 km/sec
Orbital speed

A B C D

A 5000 mph   2 km/sec
B 10,000 mph 4 km/sec
C 17,000 mph 7 km/sec
D 25,000 mph 11 km/sec

Launching spacecraft from Earth

All the rockets in the diagram are being pulled down by Earth's gravity. A and B fall back to Earth, C is going just fast enough for its 'fall' to keep it in orbit. D has enough speed to escape from Earth's gravity altogether.

Saturn V - the rocket that sent astronauts to the Moon. The first stage, 42 metres (138 ft) high, carried 2000 tonnes of petrol and liquid oxygen, which all burned in just 2.5 minutes.

The second stage took over at an altitude of about 62 km (38 miles), boosted the spacecraft to 185 km (115 miles) and almost to orbital velocity.

These are the gigantic engines of the Saturn V rocket

The third-stage rocket after release of the lunar module. All fuel tanks have by now been jettisoned.

THE SPACE ELEVATOR

Rockets are immensely expensive, and also inherently dangerous; since the vast quantities of fuel they carry turn them into potential bombs. There is, in principle at least, another way of getting into space – by means of a space elevator. After ideas floated by Polish-Russian rocket scientist Konstantin Tsiolkovsky in 1895, Russian engineer Yuri Artsutanov in 1960 and American physicist Jerome Pearson in 1975, the British writer Arthur C Clarke described a space elevator in detail in his 1978 novel *Fountains of Paradise*.

From a tower standing 50-metres (165-feet) high on the equator a cable stretches up into space and is tethered to a large mass – perhaps an asteroid – so that its centre of mass is in a geostationary orbit, 36,000 km (22,000 miles) above Earth's surface. In effect the cable would then be in orbit around Earth. Lifts, or elevators, could then run directly up the cable and into space. Powered by electromagnetic induction, like maglev trains, these would have no wheels or other moving mechanical parts, and would be able to travel at immense speeds, perhaps thousands of kilometres per hour. The advantage of such a system would be its simplicity and its cheapness: once it was up and running it would cost thousands of times less per kilogram of payload than conventional rockets. Mass space tourism might even become possible, with a trip into orbit costing no more than a flight to Ibiza.

This was only science fiction until a few years ago, because there was no material strong enough for the cable. Now, however, scientists are able to make carbon nanotubes, which are not only very

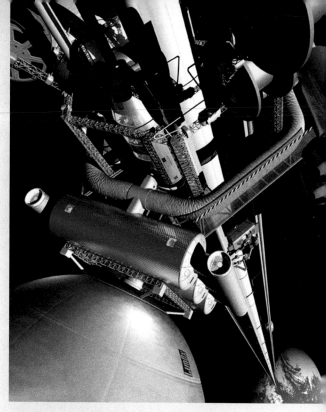

A NASA artwork envisages a space elevator lifting payloads with a pulley system.

light but 100 times stronger than steel. Today there are regular workshops to discuss the construction of a space elevator. Sir Arthur Clarke himself was asked when it might become a reality. He replied, 'Probably about 50 years after everybody quits laughing.'

STAYING IN ORBIT

People are sometimes puzzled about how spacecraft can stay in Earth orbit without falling down. Adam likes to think of himself standing on a high mountain with a handful of cricket balls. If he drops one from head height it falls to Earth in a little less than half a second. If he throws one away from him horizontally it will still fall to Earth at the same rate, but it will travel some distance, and fall much further, because the ground is lower than his mountain top. The harder he throws a ball, the further it will travel, until if he could throw it hard enough, it would travel all the way round the world and hit him on the back of the head. It would be falling all the time, but travelling so fast that its curve would match the curve of the Earth, and it would stay at the same altitude.

This is impossible near Earth's surface, because the drag from the air would slow the ball down, but if the ball started say 300 km (190 miles) up, where there is no atmosphere, then it is perfectly possible, and that is indeed how satellites and other spacecraft stay in orbit. If you tie a weight on the end of a piece of string and whirl it around your head, you have to pull on the string to keep the weight in orbit; the pull of Earth's gravity keeps the spacecraft in orbit. Similarly the pull of Earth's gravity keeps the Moon in orbit, and the pull of the Sun's gravity keeps all the planets in orbit.

From the Apollo missions we learned a good deal about the Moon. Moon rock was brought back to Earth for examination, and the final mission, Apollo 17, included a geologist, Jack Schmitt, the first scientist to go into space. In reality, however, the Apollo project was primarily for publicity; as Kennedy had said, 'No single space project in this period will be more impressive to mankind...' At the time, Ladbrokes betting shops offered odds of 10,000 to 1 against Kennedy's assertion coming true.

Altogether 27 astronauts left Earth orbit on Moon missions, but no one else has been so far away; all other human spaceflights have been confined to orbiting Earth. In 1973 the Saturn V rockets were used to raise Skylab into orbit. This was a small workshop where the crew stayed for a month or two, and in 1975 Americans and Russians collaborated in the Apollo-Soyuz missions, where the separate spacecraft docked, and the crews worked together for a couple of days. The Russians launched the first Salyut space station, Salyut 1, in 1971; the last, Salyut 7, was in orbit for nine years, coming down in 1991. They followed those with the Mir space station, which went up in 1986 and came down in 2001, after entertaining a total of 137 cosmonauts. Finally, in 2000, the International Space Station was occupied for the first time.

DATA FILE: GEOSTATIONARY ORBITS

Satellites in Earth orbit take different times to go around Earth. In low Earth orbit – say 300 km (190 miles) above the surface – the revolution period is about 90 minutes, but this gets longer for satellites further out. At 36,000 kilometres (22,000 miles) the period is just 24 hours. Because an object in a geostationary orbit takes exactly 24 hours to orbit Earth, it stays exactly 'fixed' above one point on the ground – as long as that point is on the equator. Television satellites are put into geostationary orbits above the equator so that antenna dishes can keep pointing in one direction – which in the northern hemisphere is south – towards the equator.

The crews of the 1975 Apollo-Soyuz mission pose proudly with models of the two docked spacecraft.

THE INTERNATIONAL SPACE STATION

Work began on the International Space Station in 1998, and is scheduled to be completed in 2010. In theory it will be occupied permanently by a succession of crews from several countries. It is supported by the space agencies of the US, Russia, Japan, Canada, Europe, Brazil and Italy. The early crew members were American and Russian, but altogether astronauts from 14 countries have so far visited the station. There have also been tourist visits, a wedding in space, and a golf-ball shot around the world.

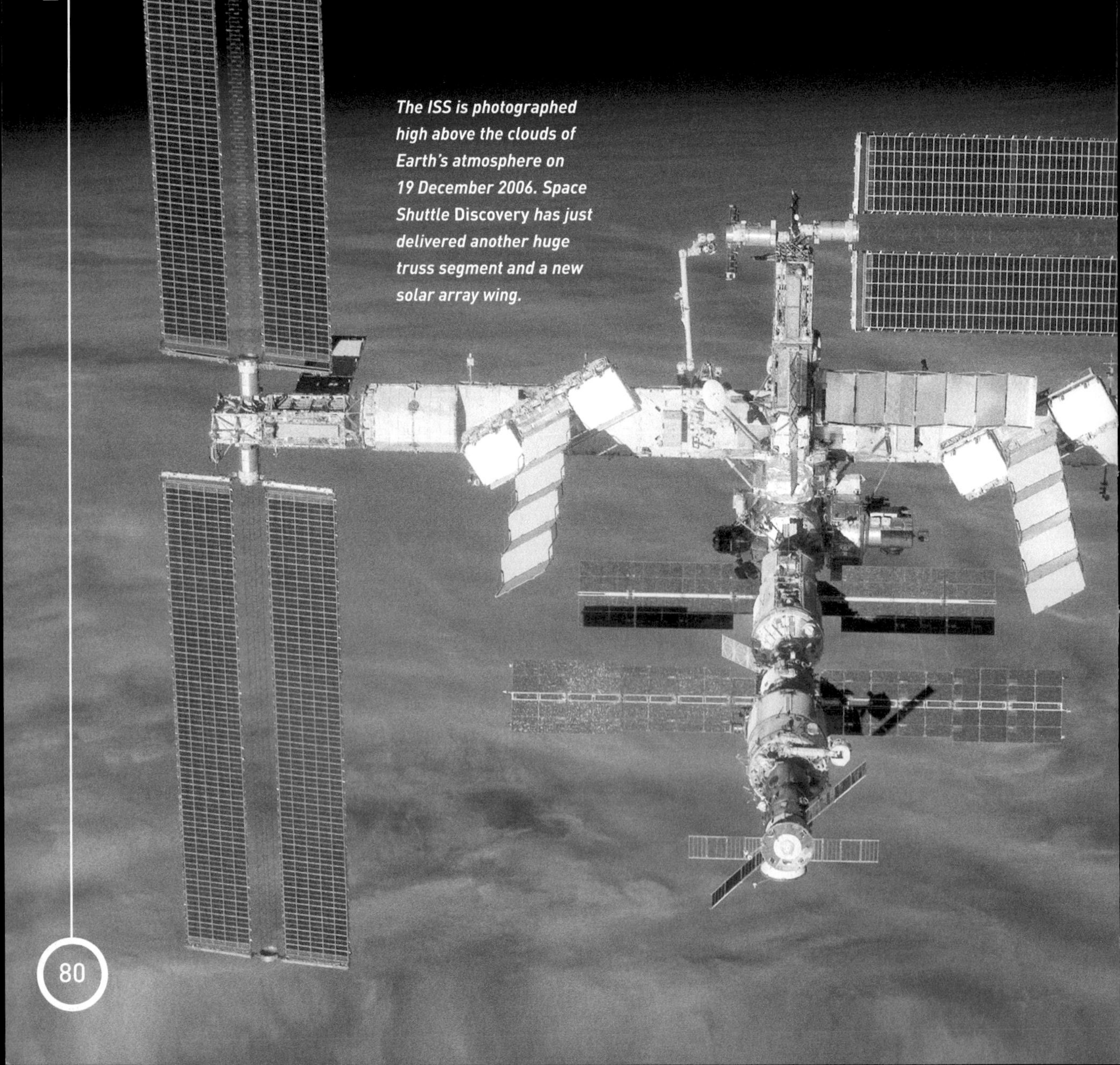

The ISS is photographed high above the clouds of Earth's atmosphere on 19 December 2006. Space Shuttle Discovery has just delivered another huge truss segment and a new solar array wing.

The ISS is in orbit 350 km (220 miles) above Earth, and has an orbital period of 92 minutes, travelling at 28,000 kph (17,000 mph), or 7.8 km (4.8 miles) per second. It is 45 metres (150 feet) long, 73 metres (240 feet) wide – across the solar panels that provide power – and 28 metres (92 feet) high. It will have a mass of 450 tonnes, and the interior will have as much space as a Boeing 747. Travel to and from the ISS is by Russian Soyuz and Progress spacecraft and by the US Space Shuttle.

The construction of the ISS was seriously delayed by the *Columbia* disaster of 1 February 2003, when, after a 16-day scientific mission, the spacecraft was destroyed on re-entry, killing the entire crew of seven. The shuttle programme was halted for two and a half years.

In 2001 the Leonardo Multi-Purpose Logistics Module arrived from Earth in the payload bay of Space Shuttle Discovery. *The ISS, when completed, will essentially be made of a set of communicating pressurized modules connected to a truss. Currently the ISS comprises four main modules.*

A 2007 shuttle mission will bring the European module Columbus to the ISS. Here's Adam having a nose around a replica of the module.

A 'space window', the cupola, will be attached to the ISS in 2010. The window will afford spectacular views of Earth.

Two blazing SRBs carry Space Shuttle Atlantis off the launch pad. Five shuttles were built, of which three remain. The winged shuttle orbiter is launched vertically, usually carrying five to seven astronauts.

THE SPACE SHUTTLE

The space shuttle is the successor to the Saturn V rocket, and most of it is recycled from earlier missions. The crew live and work in the main body of the shuttle, the orbiter, which docks with the International Space Station; after each flight it glides back to Earth for a landing on a long runway.

The two solid rocket boosters (SRBs) deliver 80 per cent of the take-off power. Each contains nearly 500 tonnes of fuel – aluminium powder mixed with the oxidant ammonium perchlorate. They burn for the first two minutes of flight, then, at an altitude of 45 km (30 miles), they separate from the orbiter and parachute down into the Atlantic to be collected and reused. The rust-coloured External Tank feeds half a million gallons of fuel (liquid hydrogen and liquid oxygen) to the three main engines during the launch, and is jettisoned at an altitude of 113 km (70 miles), after a burn of 8.5 minutes, by which time the orbiter is at orbital speed of 27,000 kph (17,000 mph) or 7.5 km (4.6 miles) per second.

SPACE SCIENCE

What do astronauts do in space stations? They spend part of the time just living – sleeping, cooking, eating, exercising – and much of the rest of the time working. There are instruments to maintain and repair; there are physiological checks to carry out to make sure they are healthy; and there are scientific experiments to conduct.

Some of these are in the form of demonstrations. At the end of the Apollo 15 mission, David Scott, standing on the Moon, dropped a hammer and a feather at the same time. On Earth the feather would fall slowly, because of air resistance. But on the Moon, where there is no atmosphere, they fell together, and hit the surface at the same time, showing that the weight does not affect the rate at which things fall as long as they are not affected differently by the air they fall through. This was what Galileo had said in 1590, and what Adam demonstrated a few years later by dropping tomatoes of different sizes from the Leaning Tower of Pisa. NASA has an extensive life sciences research programme in which astronauts investigate a whole range of things, from the effect of weightlessness on their own bodies to how plants and animals grow in space. For plants and small animals, the scientists want to follow development through at least two complete generations, to see whether micro- or zero-gravity affects their lives. There seem to be no problems with invertebrates; so next will be small mammals and fish.

There are many answers being sought. What effect does weightlessness have on the inner-ear sensors and the brain? On dexterity? On reaction to stimuli? Do small animals experience the same loss of muscle and bone as humans? Can plants grow normally without gravity? Which parts are gravity-sensitive? Can enough food plants be grown to provide food for a crew of astronauts? The Apollo-Soyuz missions, Skylab and the later space stations forced the planners to confront the problems and challenges of both short and prolonged spaceflight. First the astronauts had to cope with the high-g forces caused by the acceleration of take-off, which turned out not to be too much of a problem provided the astronauts were well-supported and strapped in.

DATA FILE: SPACE EXPERIMENTS

This scientist on board the ISS is monitorinig the growth of plants in zero gravity. There have also been space experiments in growing protein crystals. Free from the distorting influence of gravity, scientists hope to get a better understanding of protein structure. Tissue cultures may behave differently when the cells are weightless. And observation of Earth from space should provide useful information about global warming, desertification, and so on.

Astronaut David Scott performs his hammer and feather demonstration on the Moon for the benefit of millions of television viewers.

WEIGHTLESSNESS

Then there is the question of weightlessness, or zero-gravity. In Earth orbit, astronauts are not really weightless, nor is gravity zero. What is really happening is that the spacecraft is continuously falling towards the Earth, and so is everything inside it. Because they are all in free fall together, there is no resultant movement or force between the astronaut and the 'floor' of the spacecraft. You can get the same effect while bouncing on a trampoline: all the time you are in the air you are in free fall, or weightless. Imagine trying to pour water from one (preferably plastic) container into another while you are in the air; you would find it rather difficult. It's commonly called weightlessness or zero-gravity, because that is what it feels like to the astronauts.

Crew on board the ISS enjoy a moment of zero-gravity bonding. Although fun to begin with, weightlessness has some serious drawbacks.

In the zero-gravity environment, things tend to float about unless they are fastened down. You can't just walk across the 'floor'; you need Velcro or magnetic boots, or something similar. Nor can you put things on a table, or on shelves: everything has to be restrained. Weightlessness often causes space sickness, which, like sea-sickness, is brought on by the conflicting signals sent to the brain about up and down. The semicircular canals in your ears work (rather slowly) by gravity, while your eyes tell you where the horizon is, and therefore which way is up and which way is down. In zero-gravity, the ears are hopelessly confused, and the eyes often have no clear horizon; the brain reacts by making you feel sick. Many astronauts have experienced severe space-sickness, but luckily it seems to pass within two or three days. Lack of exercise for long periods causes some wasting of the muscles, and to minimize this the space stations are equipped with exercise bikes, vibrating platforms and other machines, modified so that the astronauts can be bungeed on.

There is also a reduction in blood volume after some time in zero-gravity, which is fine while you are up there, but combined with weakened muscles and recovered weight it can make astronauts feel most peculiar and unsteady when they return to Earth. This could be a problem on a long trip to Mars, because they would need to be at the peak of alertness and fitness when they arrive. Because the limbs have to do much less work supporting the weight of the body, the system decides the skeleton does not need to be so strong, and so there is considerable loss of calcium from the bones; this can be as much as 1 per cent every month. After prolonged stays in space this may present a serious problem.

MEDICAL EMERGENCIES

When astronauts go on longer and longer missions, there are increasing chances that one of them will get ill, or be injured, or simply get a terrible toothache. Most missions will therefore include a doctor capable of dealing with such emergencies, and all astronauts have to be trained in first aid – and indeed have to be multi-skilled, so that they could take over the tasks of anyone who is incapacitated. There also arises the difficult question of what to do if someone were to die in space. The rest cannot carry a body around with them for weeks; would they simply 'bury' their colleague out in the void?

RADIATION

There is increasing danger from radiation as you get further from Earth. Here on the surface we are protected by the ozone in the atmosphere which cuts out most of the ultraviolet rays from the Sun, and also by Earth's magnetic field, which turns away most of the harmful cosmic rays. In orbit around Earth you would be more at risk, although the ultraviolet would not get through the walls of the spaceship. If astronauts were to visit Mars, ultraviolet radiation might

DATA FILE: ARTIFICIAL GRAVITY

Scientists have dreamed up various ways of generating artificial gravity. In movies we've seen ring-shaped sections of spacecraft rotating, creating artificial gravity at the outside (above), but in practice this is rather complex. One solution might be to make a spacecraft in two parts, one perhaps containing the engines, the other comprising the living quarters; tie these together with a cable, and then set them spinning around their common centre of gravity.

Space-station crew must exercise for several hours a day to avoid muscle wastage and bone loss.

be a more serious problem. Mars has no ozone layer, and the surface is bathed in ultraviolet rays all day. Spacesuits would have to protect against this, or astronauts would get seriously burned.

Cosmic rays are more serious than ultraviolet, and are best stopped by material containing lots of hydrogen atoms. Curiously, polyethylene, one of the simplest of polymers, turns out to be as good as anything, and a 30-cm (12-inch) wall of polyethylene around the spacecraft would absorb 30 per cent of the radiation. The rest will get through, however, and one way to minimize its effect is to feed the astronauts plenty of vitamin C and other foods high in antioxidants, which will mop up the harmful ions and free radicals generated by the radiation.

SPACE FOOD

Food and drink need some care. In zero gravity, food will not stay on a plate, nor drink in a glass. In the early days everything came in aluminium tubes: cosmonaut German Titov's meal in 1961 was a tube of vegetable puree soup, a tube of liver pate, and then a tube of blackcurrant juice. Today the food is much more appetizing, whether it is Russian and canned, or American and dehydrated. There is a wide variety, and the astronauts eat it with knives, forks, and spoons that are magnetized to keep them on the table, or on a tray strapped to the knee.

The food itself comes in cans or sachets, and, after rehydration it is sticky enough to stay on the spoon or fork. Drink is kept in plastic squeezy bottles, and delivered straight into the mouth. The astronauts have to be careful not to spill anything, since crumbs and drops of liquid floating around the cabin are definitely bad news: they clog up the instruments and the air filters, and can get in the astronauts' eyes. For some reason astronauts generally eat too little while they are in space, and often return 5 per cent lighter. In extreme cases this can be serious, and astronauts seen to be losing body mass are encouraged to eat more. You can try space food yourself. Several companies offer to provide exactly the same food as is prepared for the astronauts, and claim it to be ideal for camping trips.

No one ever associated hamburgers with losing weight before. Here a dinner floats free of its diner aboard the ISS.

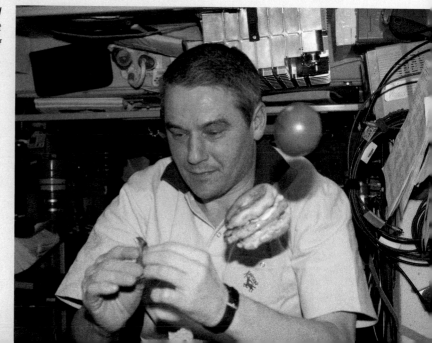

THE ZERO-GRAVITY LAVATORY

Lavatories have continued to be problematic ever since Alan Shepard's first flight. In the early days astronauts wore nappies ('intimate contact devices' in NASA's elephantine euphemism), but these were no longer acceptable in flights of longer than a few hours. For Skylab there was a new system of waste management – the Waste Collection System or WCS, with a modesty curtain to shield the user from view.

The WCS comprised a cylinder about 80 cm (50 inches) high and 30 cm (12 inches) across, like an old-fashioned spin dryer. From the front came a flexible plastic hose, as from a vacuum cleaner. This urine collector was intended for both men and women, and was fitted with a triangular rubber nozzle on the end. Unfortunately the sexes have different peeing systems, and as a result the nozzle never fitted anyone very well. What's more, the vacuum was never much good, so the nozzle was always a bit wet from the last user.

When you wanted to poo you took your trousers down and sat on top of the cylinder, with your feet in stirrups, and then pulled a pair of spring-loaded restraints over your thighs. Remember, you were weightless, and you would not want to float off in the middle of the operation. Then you opened the sliding lid and performed. Next problem, zero-gravity. On Earth, the poo just falls, but it is sticky, and in zero-gravity it just stays put. So the seat was fitted with eleven channels to blow air upwards from all around so as to peel the poo away from you and into the cylinder. Unfortunately, the air was icy cold. Once inside the cylinder the poo was spun to the outside and freeze-dried to keep it out of the way. This WCS was definitely a challenge, and at least one astronaut ate nothing at all for an entire mission in order to avoid having to use it. Sadly this does not work. The body produces solid waste even when it receives no food, so he went horribly hungry for no good reason.

PSYCHOLOGICAL CHALLENGES

There is also a range of mental stresses associated with spaceflight. First there is the claustrophobia of being shut in a metal box for an extended period of time. The Apollo astronauts, who travelled in threes, were in such close quarters they were almost touching one another all

This prototype zero-gravity lavatory is similar to the WCS toilet. Air flow is used to suck waste into a disposable container where it is dried. Unfortunately, in the WCS toilet the poo became so dry that bits began to flake off and drift around the cabin. This was a pity, because the crew were in the habit of flicking peanuts to one another, and eventually the only way they could tell what was what was by the taste.

'It's a high-stress environment – low gravity, being confined; you have to learn to put up with people for weeks to months on end'

Nick Kanas
space psychologist

the time, and could scarcely change places or stretch their limbs. In the ISS there is more room, but you cannot just nip outside for a minute's peace and solitude. You are also in close quarters with several other people, and cannot escape from any of them. You have to hope that none of them has bad breath or smelly feet – and if they have then you just have to put up with it, day and night, for the entire mission. One of your fellow astronauts may have a loud voice, or a raucous laugh, or a habit of talking all the time, or belching, or not listening to you, or any number of other little habits that over a long time might be hard to live with.

You might feel lonely and isolated from your family and friends; you might get bored and depressed, although in practice, depression does not seem to be nearly as much of a problem as it is with submariners. Astronauts seem to get enough support from ground control to keep them from getting depressed. Boredom can be alleviated by DVDs of movies and games. However, contact with family, although easy from orbit by phone, becomes impossible on longer missions. Even from the Moon there would be a three-second delay in exchanges, which makes conversation extremely difficult – by the time you hear the answer to one question you have thought of two more questions – and from Mars the delay would be about three-quarters of an hour. Communication would have to be by email.

There are some problems that you might not easily think of in advance. One is peculiar smells. A few years ago some Velcro was sent up to the ISS for fastening things out of harm's way, but in those confined quarters it had a distinct smell, and left a nasty taste at the back of the throat; so they sent it back on the next shuttle – but even that could be weeks away. The trouble is that you can't just wind the window down for a breath of fresh air.

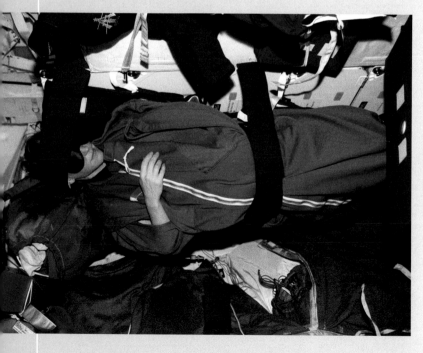

Sleep is quite difficult when you cannot just lie in a bed. The astronauts sleep in tethered sleeping bags, or hammock bags, often tied to a wall in a vertical position, relative to the 'floor', since that uses less space – and it makes no difference to the body. In some spacecraft there are bunks, in which astronauts can sleep either 'on top of' or 'underneath' the boards – hard boards feel soft in zero gravity. On average, astronauts sleep for one hour less in space than they do on Earth.

NASA's bloodhound is George Aldrich, who has been smell-testing materials for years – at least 700 objects so far, and counting.

Like anyone else under stress, astronauts tend to pass frustration on to innocent parties. Just as at home, after a bad day in the office being shouted at by your boss, you might go home and shout at your partner, or kick the cat, so astronauts, frustrated in a spacecraft, may shout at Mission Control, or, worse, take it out on their families when they get home.

Russians and Americans behave quite differently in space, and if a mission has people of various cultural backgrounds this may lead to unexpected tensions, not to mention misunderstandings. On the other hand a multicultural group can be highly effective. The A380 Airbus is built in Toulouse by a large team from France, Germany, Spain, Britain and other countries. The team leaders report that they all have the same sort of engineering training, but they approach problems from different directions, which is often effective.

No crew should have a small minority in case someone is isolated, picked on or teased; this happened on Mir when one American joined two Russians. So no crew should have one American and six Russians, or one woman and eight men. Ideally there should be roughly equal numbers of men and women, and of each ethnic group, and a total crew size of perhaps seven; an odd number is good for making decisions by majority vote.

EXTRA-VEHICULAR ACTIVITY

Astronauts sometimes leave the spacecraft for space walks, or EVA, and many describe this as the best part of space travel. They have to wear pressurized space suits and carry an independent life-support system, breathing pure oxygen, and the space suit is bulky and rather clumsy. In particular the gloves are heavy and stiff, and after a few hours can make the hands tired and sore – it's like squeezing a rubber ball repeatedly for hours. However, there is an extraordinary sense of freedom, like being in a swimming pool but without the water to push through. The view is breath-taking – you can see half of Europe at the

BIOG FILE: NICK KANAS

Nick Kanas, Professor of Psychology at the University of California at San Francisco, has been studying the mental health of astronauts in space for 15 years. He advises on human behaviour for the Committee On Space Biology and Medicine Discipline. In particular he has studied group behaviour in both the NASA and Mir crews – how they interact with one another, with the ground crews and with their families.

'For some astronauts it was very difficult for them to readjust just when they got back'

Nick Kanas
space psychologist

same time, and days and nights are only 45 minutes long. Unfortunately they aren't out there just to enjoy the view: there is always work to be done – especially repairs to the spacecraft, to an artificial satellite, or to the Hubble Space Telescope (see pages 54–5).

THE FUTURE OF HUMAN SPACE EXPLORATION

Going to the Moon was the obvious first step in the human exploration of the cosmos; the obvious next step is Mars, which is further away than Venus but very much more hospitable. People have always wanted to go to Mars, and to find out more about the Red Planet – but is it possible?

First, getting to Mars would take at least eight months, the trip back would take as long again, and astronauts would want to spend months there; so a total trip of two or three years is necessary. For, say, seven astronauts, the sheer mass of food, water, oxygen and other life-support materials needed for that length of time is vast. Even the Saturn V could lift only a fraction of it into space. The spacecraft and the life-support materials would have to be assembled in space, like the ISS.

Second, no human being has stayed in space for anything like such a long time. American astronaut Sharon Lucid spent 188 days – six months – in the

American astronaut Sunita L Williams participates in an EVA session on the ISS in January 2007. During her seven-hour space walk, Williams reconfigured cooling loops for one of the modules, rearranged electrical connections and secured the starboard radiator of the P6 truss after retraction.

Russian space station Mir, and wrote home that every Sunday she wore pink socks and shared a bowl of jelly with the two Russian cosmonauts. To the surprise of the experts, she was able to walk from the shuttle when she finally came down. The record for the longest continuous stay in space is held by Russian cosmonaut Dr Valeriy Polyakov, who spent 438 days on the Mir space station, from January 1994 to March 1995. In space he exercised for between 90 minutes and three hours every day, and when he landed he was able to walk to a nearby chair. The next day he went jogging. From these experiences it seems that astronauts could spend eight months travelling to Mars and then be able to walk on the planet, but what state would they be in by the time they got back to Earth?

Third, human beings have short operational lives. Even the three years of a Mars expedition – an extremely long time for a human journey – is barely more than a tenth of the Voyager missions, which are still working after 29 years.

Fourth, the cost would be astronomical. As part of Project Aurora, the European Space Agency plans a series of missions to Mars, culminating with a manned mission in 2030. NASA has also hinted at a manned mission to Mars, but nothing is yet settled. British sports scientist Adam Hawkey reckons that if they do get there, astronauts will run rather than walk on Mars. Martian gravity is only about one-third of Earth gravity, and with a spring in your step, running will be more efficient than walking.

In 1998 American aerospace engineer Robert Zubrin and others founded the Mars Society, an organization devoted to selling the idea of human exploration and colonization of Mars. The first step – Mars Direct – would be to send an unmanned Earth Return Vehicle, which would go direct to Mars and deliver a small nuclear reactor, a small chemical factory and a supply of hydrogen, which could be combined with the carbon dioxide in the Martian atmosphere to make methane and oxygen. This would provide the necessary fuel for the astronauts, who would travel on a subsequent journey, along with a Mars Habitat Unit (MHU).

The Mars Society also runs Mars Analog Research Stations (MARSs) to try out prototypes of the MHU in Mars-like conditions. One MARS, in the

Utah desert, is a circular two-floored building with sleeping and living quarters for six upstairs and an open-plan working area below. There is an 'airlock' in which the participants spend five minutes to 'depressurize' before going out into the desert for EVA. There, wearing a full spacesuit, complete with gloves, helmet and backpack, they carry out the sorts of work that astronauts would – collecting rock samples, searching for signs of water, and so on. They also hope to find out about group dynamics: what is the best number of people and how they get on after two or three weeks confined in the same building.

Does it make sense for humans to go into space? They are far better than robots at using their intelligence, at making decisions, at spotting potential problems and at putting things right. On the other hand they are enormously expensive in terms of life support: they eat masses of food, drink masses of water, breathe masses of air, and they need to sleep for several hours every day. Robots have none of these disadvantages. Buzz Aldrin, who set foot on the Moon, is sure that humans should

'boldly go…'; when asked why, he said he was sure the Creator would have wanted us to do so.

PROBE AND ROBOT MISSIONS

Since that first Sputnik in 1957, some 200 space probes have been launched, with varying degrees of success. In the early years both Russian and American rockets were rather unreliable: almost half the launches failed, but the technology gradually improved and spectacular results were achieved.

One dramatic event happened on 3 February 1966, when the Russians managed to land Luna 9 on the surface of the Moon, complete with a television camera to beam back photographs. This was the first ever soft landing on the Moon – another spectacular first for the Russians, even though it was their twelfth attempt. At Jodrell Bank near Manchester, England, British physicist Bernard Lovell heard about this, and immediately swung his great radio telescope towards the Moon to see whether he could pick up any signals. Sure enough, he found a strong chatter of what appeared to be

NASA's Mars Exploration Spirit *captured this stunning view of the Sun setting below the rim of Gusev crater.*

some sort of code. Reasoning that it might be pictures, he sent the signals directly to the *Daily Express*, who fed them into their standard picture-receiving machine, and out came the first ever close-up photographs of the surface of the Moon, which were splashed across the front page the following morning. The Russians were apparently a little miffed both by the fact that their pictures appeared first in a decadent Western newspaper, and also by the fact that the horizontal scale had been shrunk in the process, so that the pictures were somewhat distorted. Or perhaps they had deliberately used standard camera equipment, knowing that Jodrell Bank would produce pictures of higher resolution than any Russian telescope.

There are four main plans for missions to planets:

1. Flyby – to take photographs and measure magnetic fields;

2. Orbiter – to go into orbit around the planet, which gives much more time to collect detailed information;

3. Lander – to land a craft gently on the surface so that it can send back photographs and

Radar mapping of Venus reveals it to be a world of canyons and volcanic mountains. Thick, poisonous clouds cover the surface, where temperatures are hot enough to melt lead.

chemical analyses, and measure temperature, wind speeds, and so on;

4. Crash landing – to get some idea of the composition of the surface and the soil beneath.

Of these four, the flyby and the crash landing are the least likely to go wrong. Going into orbit around a distant planet requires some clever and accurate calculations, while landing presents all sorts of difficulties. On Mars, for example, the atmosphere is so thin that parachutes are hardly enough to effect a gentle touchdown. There is always the danger of landing on the side of a rock or cliff, or in a lake; any of these would probably be the end of the mission. The first British probe, Beagle 2, came to a sticky end on Mars; it was supposed to land on 25 December 2003, but never phoned home, and no one knows whether it broke on impact or fell into a ravine.

Both the Russians and the Americans launched a variety of probes both to the Moon and towards the planets. Pioneer 10 flew past Jupiter in 1973; Pioneer 11 flew past Jupiter and Saturn. Vikings 1 and 2 went to Mars, and several probes went to Venus; Magellan mapped the Venusian surface by radar, since the thick clouds prevent any visual contact. The first free-ranging robotic vehicle to wander about on Mars, Sojourner, was delivered there by Mars Pathfinder in 1997, and between them they eventually sent back 17,000 pictures as well as chemical analyses of the soil.

93

MARS PROBES

Since the Viking landers of the 1970s, NASA has put three mobile robotic probes on Mars and more are to follow. Mars Exploration Rovers *Spirit* and *Opportunity*, launched in 2003, are still trundling across the planet and sending back holiday snaps. They were expected to work only for a few months, because they are solar-powered. The designers thought that the solar panels would soon get covered with dust that would block the sunlight and cut off the power. Luckily it turns out that the Martian winds blow the dust off, and they have now exceeded their planned lifetimes by 12 times, and have

Seen here for comparison are models of the first mobile robotic probe, Sojourner *(left), and the most recent, a Mars Exploration Rover (right).*

NASA's Mars Science Lab will be the size of a small car. It will specifically look for organic material in the soil – evidence of life or past life.

sent back about 170,000 pictures, which have greatly increased our knowledge of the Red Planet.

Next in line is Phoenix, scheduled to launch in 2007, and headed for the Martian north pole to search for organic material after the ice has retreated in the spring. In 2009 NASA's Mars Science Lab is scheduled to set off. This is an entire laboratory, powered by plutonium, which should ensure a working lifetime of many years.

Other remoter possibilities are to fly balloons or aircraft above the Martian surface. This would allow rapid surveying of a large area, but would be difficult because the atmosphere is so thin – only about one-hundredth of the pressure of Earth's atmosphere – that getting enough lift would be tricky, and the payload would have to be small.

DATA FILE: EXOMARS

In 2012 the European Space Agency hope to launch Exomars, which will drill a few metres below the surface to look for organic material. In addition to spectrometers and so on, Exomars will use biology to look for biology: protein receptors will actively seek out special molecules associated with life. The instrument comprises a series of small chambers, each a few millimetres across and containing proteins that will latch onto molecules of a specific type. The machine will dissolve a chunk of rock in a solvent, label the solution with a dye, and float it across the chip.

The Mars Exploration Rovers' mission is to examine rocks and soil and to search for evidence of Mars's warmer, wetter past.

One of Phoenix's missions will be to look for a habitable zone that may exist in the ice-soil boundary near the planet's poles.

Several probes have flown past Saturn, giving us wonderful pictures of its rings, and the latest Cassini-Huygens double-act has provided a mass of new data about two of Saturn's moons – see pages 100–1.

Comets and asteroids are interesting, because they may be able to tell us something about how the planets were formed (see pages 33–5). Giotto, the first deep-space mission of the European Space Agency followed five other probes – Soviet, Japanese, and American – to fly by Halley's comet in 1985, passing within 600 km (370 miles) of the core on 13 March. During its close encounter, Giotto was hit by some 12,000 pieces of high-speed dust, one of which, with a mass of perhaps a gram, sent the half-tonne spacecraft spinning out of control for half an hour before radio contact could be restored. Giotto found that rather than being a dirty snowball, the comet is made mainly of dust , with embedded ice. The dust is very dark, rich in hydrogen, carbon, oxygen and nitrogen, with some minerals and organic compounds, and was being knocked off the comet by the solar wind at a rate of about three tonnes per second.

The NEAR-Shoemaker probe went into orbit around asteroid 433 Eros in 2001, and sent back useful information. The Japanese probe Hayabusa (falcon) orbited and then touched down briefly on asteroid Itokawa in November 1985, fired a missile into the surface and collected samples of the dust produced. Landing on an such a chunk of space rock is extremely difficult, because they are so small and have essentially no gravitational pull; indeed it might be called 'docking with' rather than 'landing on' the asteroid.

VOYAGER

Two of the most impressive of the spacecraft are NASA's Voyagers 1 and 2, which have been to the outer planets. The probes took advantage of a line-up of the planets than happens only once every 176 years. As they passed each of the gas giants they made use of the slingshot effect: the huge gravitational pull of each planet flipped them on their way with a higher speed. Voyagers 1 and 2 were both launched in 1977. Powered by radioactive sources, both are still sending back information, even after 29 years.

Giotto sent back a mass of information about Halley's Comet (above) showing that the body is about 15 km (9 miles) long by 7–10 km (4–6 miles) wide.

In Victorian times rich young men were sent on grand tours of Europe for their education, and Voyager 2 (below) has made a grand tour of the outer solar system, visiting Jupiter, Saturn, Uranus and Neptune.

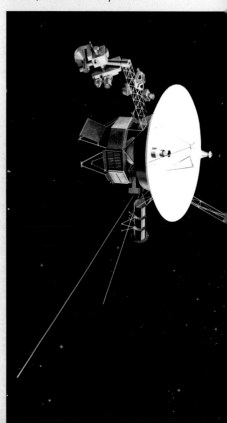

Voyager 1 passed within 350,000 km (220,000 miles) of Jupiter in March 1979, photographing its surface and its moons; then on to Saturn in November 1980, passing 125,000 km (78,000 miles) over its cloud tops, and discovering the complex structure of its rings. It also saw that Saturn's moon Titan has an atmosphere, and so the mission controllers decided to make a close flyby of Titan to gain more information. This gave Voyager 1 another slingshot flip, and sent it out of the plane of the ecliptic.

Voyager 2, launched on a slower path than Voyager 1, sailed 600,000 km (370,000 miles) above Jupiter in July 1979, showing that the Great Red Spot is a giant storm, and taking photographs of its nearest big moon Io, which revealed nine active volcanoes on the moon, with plumes rising 300 km (190 miles) into space. This volcanic activity, the first observed away from Earth, is probably caused by tidal heating; Io is so close to Jupiter and neighbouring

moons Europa and Ganymede that it is tugged to and fro by them, and its innards must be sloshing to and fro in the powerful gravitational field. Voyager 2 reached Saturn in August 1981, taking photographs of the rings; Uranus in January 1986, skimming only 80,000 km (50,000 miles) above its cloud tops, and telling us that it has a powerful magnetic field, ten more moons than we knew about, and a day length of only 17 hours; and Neptune in August 1989, where it discovered a dark spot, a huge storm like Jupiter's Great Red Spot.

Both Voyagers are now heading out of the solar system altogether. Voyager 1 is headed north, 35° out of the plane of the ecliptic, and, travelling at about 1.6 million km (1 million miles) a day will reach another star in only 40,000 years, while Voyager 2, travelling slightly more slowly, is headed 48° south of the ecliptic plane, and will take 300,000 years to reach Sirius, the brightest star in the sky. Both Voyagers are now in the heliosheath, where the Sun's influence is small. The entire solar system travels through space protected from outside influence rather like a car surrounded by a huge bubble full of air. The solar wind – a stream of particles sprayed out by the Sun – zooms through the solar system at 1.5 million kph (900,000 mph), but as the Voyagers approach the heliopause – the surface of the bubble – the solar wind will slow down, and then they will escape the Sun's influence completely.

Voyager 1, now 15 billion km (9 billion miles) or 14 light-hours away, is the most distant human artefact in the cosmos. Within ten years it should be outside the Sun's bubble, and it will then truly be on its way to the stars.

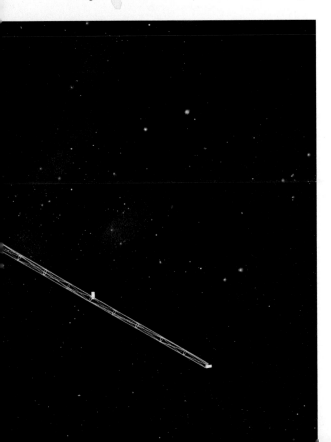

Light curve for WASP 1

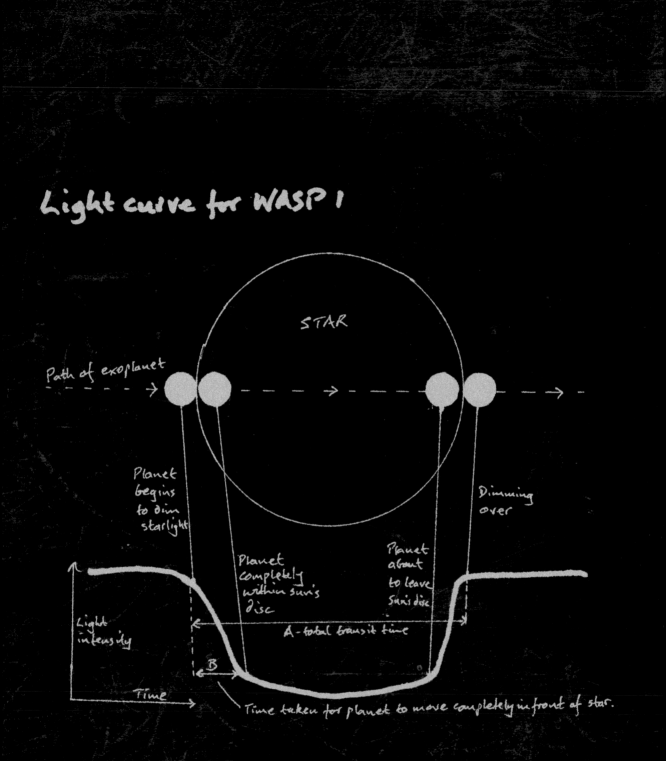

STAR

Path of exoplanet

Planet begins to dim starlight

Planet completely within sun's disc

Planet about to leave sun's disc

Dimming over

Light intensity

A – total transit time

B

Time

Time taken for planet to move completely in front of star.

4 / OTHER WORLDS

On 14 January 2005 I spent several nail-biting hours in Darmstadt, Germany, waiting to witness what we hoped would be an amazing achievement. Astronomy used to be purely an observational science, but in the past decade or three practical astrophysics has come of age. This was to be a spectacular demonstration – the landing of a spacecraft on Titan, the largest moon of the planet Saturn. I was privileged to be at ESOC – the European Space Operations Centre – to make a television programme about the event.

All morning we watched the scientists anxiously pacing up and down as the various stages were passed. The main spacecraft, Cassini, had been launched seven years earlier, in 1997, and had taken all this time to trek through the solar system and go into orbit around the ringed planet. On Christmas Day 2004 the probe, Huygens (which the Dutch pronounce 'Howgens'), separated from the mother craft, Cassini, and headed for Titan. After three more weeks the time had come for Huygens to descend through Titan's atmosphere. The battery of instruments on board had been designed, built, tested, installed and launched eight years earlier. Would they work on the day? Would they all come to life on cue and send their data back? I don't recall ever seeing scientists so tense.

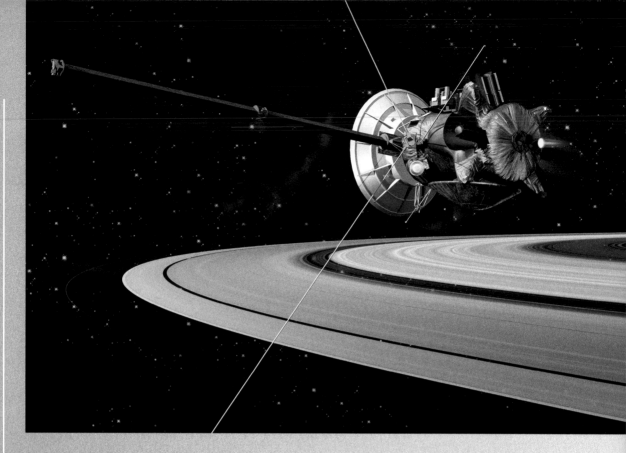

DESCENT TO TITAN

First, after descending some way into the atmosphere, Huygens had to jettison its heat shield. This would in theory activate the instruments, which would record all sorts of information about Titan's atmosphere and start transmitting it to Cassini. After it had received all the data, Cassini would swing round so that its antenna pointed towards Earth, and then relay the data to the home team. At least, that was the plan.

The signal from Huygens was extremely weak, so the Earth-bound scientists had only faint hopes that they would be able to pick it up directly. Cassini, however, had a more powerful transmitter and a bigger antenna. At 11.25 a.m. German time, the scientists gathered in the control room hoping to get the first indication. A muted cheer went up when the mighty Green Bank Telescope in West Virginia did indeed detect a faint signal from Huygens, showing that the shield had been shed and the instruments had started to work.

Then came bad news. The probe was meant to be transmitting on two channels, but only one was working; the signal to turn on the other had not been sent – or at least had not been received. Luckily the second channel was mostly redundant – a back-up of the stuff on the first – so losing it was not a complete disaster, and the first channel remained clear.

Meanwhile Huygens was gradually floating down towards Titan, deploying a succession of parachutes as the atmosphere thickened. During the descent it measured the chemistry of the atmosphere, the pressure, the temperature and the wind speed, and it took a series of photographs

This computer-generated image shows Cassini-Huygens in orbit above the rings of Saturn. The main spacecraft, Cassini, was built by NASA; the Huygens probe by the European Space Agency (ESA) and the powerful radar and antenna by the Italian Space Agency (ASI) .

DATA FILE: CASSINI-HUYGENS

The Cassini-Huygens mission was launched on 15 October 1997 to study Saturn and its moons. The entire assembly was about the size of a bus, weighing 5.6 tonnes. The craft received gravity-slingshot assists from Venus, Earth and Jupiter. It also helped to verify Einstein's theory of relativity. The theory predicts that massive objects distort space-time, and therefore that radio signals from Cassini that passed close to the Sun would show slight shifts in frequency – and this did indeed happen. Saturn is so far from the Sun that solar power is not possible, and the orbiter is powered by radioisotope thermal generators (RTGs), each with a lump of radioactive plutonium. This caused an outcry in 1997, when people feared what might happen if the craft crashed to Earth.

that showed the surface of the moon in ever-increasing detail. However, the scientists did not know this yet. They had to wait a further three anxious hours while Huygens completed its descent and sent all the information to Cassini, and then wait for Cassini to turn around and beam the data back through space before it could be received and decoded.

At last, at 5.19 p.m., the data came pouring in, and with whoops and beaming grins the waiting scientists could go to work on it. They had several years' work ahead of them simply working out the data.

From my point of view, by far the most spectacular results on the day were the photographs – the first-ever close-up surface pictures of any body further away than Mars. They showed features that looked like huge cliffs, and a great delta of what appeared to be streams or rivers flowing down into a sea or flood plain. When I saw those pictures I realized that I was looking not just at a lump of rock, but at another world – a world with cliffs and rivers and seas. What sorts of other worlds are there in the universe, and how can we find out about them?

BIOG FILE: NICOLAUS COPERNICUS

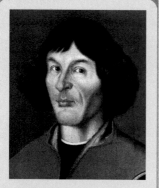

Nicolaus Copernicus was born in Poland in 1473 and educated at Cracow Academy, and at the universities of Padua and Bologna in Italy. He became a mathematician, lawyer, Catholic priest, economist and diplomat, but what he really loved was astronomy, and he made his first observations in 1497. Copernicus spent most of his working life in Frombork (Frauenburg), Poland, which is where he was buried in 1543, and where his grave was apparently found, after a long search, in August 2005. According to legend the first printed copy of his book was put into his hands just before he died.

NEIGHBOURING WORLDS

Most of the stars arc their way across the sky in repeated and predictable patterns, but the planets – the 'wanderers' – seem to have a different agenda. For most of the time they follow the same paths, but occasionally they perform little loops in the sky. To early observers this curious behaviour was obviously caused by the gods, but gradually more rational explanations were sought and found. In about 250 BC the Greek astronomer Aristarchus said that Earth and the other planets revolve around the Sun, but no one believed him. In about AD 150 Ptolemy produced a great book called *The Almagest*, describing the movements of the heavenly bodies with Earth the centre of the universe and the planets moving in little epicycles in their orbits.

This was accepted as correct until 1543, when Polish astronomer Nicolaus Copernicus published a book called *De Revolutionibus Orbium Coelestium* ('On the revolutions of the heavenly spheres'). In it he explained that Earth is not the centre of the universe; instead it revolves around the Sun, and so do the other planets. They make loops in the sky because both they and we are orbiting, and sometimes, as Earth is catching up with one of the others, it seems to go backwards in its tracks. This was powerful stuff, and Copernicus was careful not to publish the book until he was on his death-bed, lest the Church send him to the stake for heresy.

Apart from the planets and their moons, all the other points of light in the sky were regarded as 'fixed stars', until American astronomer Edwin Hubble realized (see page 14) that we are in one galaxy – the 'Milky Way' – but there are other galaxies beyond ours. We now know that there are about 100 billion (100,000,000,000) stars in the Milky Way, and at least 100 billion other galaxies in the universe.

Our own star, the Sun, has a family of planets revolving around it – a cluster of worlds we call the solar system. Are there other worlds out there – planets orbiting other stars? Since there are so many stars, the answer has always been yes, probably. But what sort of worlds might they be? Our own planets form a sort of benchmark.

Mercury, Venus, Earth and Mars are rocky planets with hard surfaces. The outer planets are gas giants; they may have rocky cores surrounded by liquid metallic hydrogen, but they have no real surfaces. They are made of gases – mainly hydrogen and helium – which are thin and wispy at the outside but get gradually hotter and thicker nearer the core. The outer planets have rings, of which the most spectacular are Saturn's. They were first seen and described by Galileo, but for 60 years no one could quite make out what they were, since as seen from Earth they looked like giant handles on the sides of the planet. The problem was solved in 1659 by the Dutch scientist Christiaan Huygens, who not only made the first

DATA FILE: PLANETS OF OUR SOLAR SYSTEM COMPARED WITH EARTH

	Astonomical unit	Diameter (000 km)	Mass	Length of day	Length of year
Mercury	0.4	5	0.06	59 days	88 days
Venus	0.7	12	0.8	243 days*	225 days
Earth	1	13	1	24 hours	1 year
Mars	1.5	7	0.1	25 hours	1.9 years
Jupiter	5	143	318	10 hours	12 years
Saturn	10	121	95	10 hours	29 years
Uranus	19	51	15	17 hours*	84 years
Neptune	30	50	16	19 hours	165 years

* Venus and Uranus spin backwards; all the rest go the same way

Astronomical unit (AU) = distance from Sun. Earth's distance from Sun = 1 AU (Astronomical Unit) = 150 million km (93 million miles); Earth's mass = 6000 million million million tonnes; Pluto, beyond Neptune, used to be considered a planet, but in 2006 was demoted to a 'dwarf planet'.

BIOG FILE: WILLIAM HERSCHEL

Frederick William Herschel was born in 1732. He came to England from Hanover, Germany, in 1755, practised as a musician in Newcastle and then got a job as an organist in Halifax. While he was there he became friendly with John Michell, a scientist who had resigned his seat at Cambridge to become a vicar. Michell was the first person to have the idea of a black hole, and described it in a letter to the Royal Society in 1783. He may have lent or given a telescope to Herschel, who soon became hooked on astronomy. When Herschel moved to Bath he began making his own telescopes, which at the time were among the best in the world. Using them, and with the help of his sister Caroline, he became famous after his discovery of Uranus in 1781. He was appointed The King's Astronomer, and died in 1822.

pendulum clock and suggested that light comes in waves, but also built a better telescope, and was able to make out that these 'handles' were actually rings going right round the planet. He also discovered Titan, on 25 March 1655. The Italian astronomer Giovanni Domenico (later Jean-Dominique) Cassini, who worked with Huygens for a time, discovered four more of Saturn's moons, and observed a large gap in Saturn's rings: he saw that they formed a double ring, one inside the other. This gap is now called the Cassini Division, and the recent mission to Saturn and Titan was called the Cassini-Huygens mission in honour of those pioneering astronomers. Photographs taken from nearby by various spacecraft have shown that there are a multitude of concentric rings, and that the rings are made mainly of small chunks of ice, all held in a narrow belt around the planet. Although the rings are 250,000 km (155,000 miles) wide they are less than 1 km (0.6 mile) thick – in fact perhaps only a few tens of metres thick.

HOW THE PLANETS WERE DISCOVERED

Five planets – Mercury, Venus, Mars, Jupiter and Saturn – are visible to the naked eye and were known to the ancients, who realized that they were not like most of the stars because of their peculiar wanderings. Copernicus listed them in the right order according to their distance from the Sun, and assumed their orbits to be circular. The German astronomer Johannes Kepler, working with more accurate observational data, showed that the orbits are not circles but ellipses.

On 13 March 1781, from his back garden in Bath, the German-born musician-turned-astronomer William Herschel discovered a new object in the constellation of Gemini. At first he thought it was a comet, but he could only observe it for a few weeks, because as spring turned into summer his new discovery appeared closer and closer to the Sun, until he could see it no longer. He had to wait until the autumn to see it again, and then realized it was not a comet, but a planet, the first new one to have been spotted for at least 2000 years. He wanted to call it *Georgium sidus* – 'George's Star' – after King George III of England, but it came to be called Uranus.

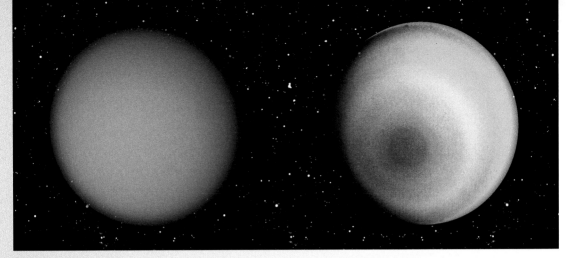

An image of Uranus was taken by Voyager 2 as it passed the planet in 1986. The natural colour (left) shows the layers of methane and photochemical smog that cover the planet. The colour-enhanced image (right) shows the concentration of smog at the south pole.

Herschel believed that most of the planets in the solar system were inhabited, but unfortunately he was a bit over-optimistic. We now know that life can exist only within limits (see chapter 6): it needs water; it needs a moderate temperature, between perhaps -50°C (-58°F) and +150°C (300°F); and it needs some protection from dangerous radiation, which might be provided by an atmosphere.

Giant planets such as Jupiter have such powerful gravity that no large animal would be able to stand up in it, even if Jupiter had any surface to stand on. Tiny planets such as Mercury have too little gravity to hold on to an atmosphere. In their search for extraterrestrial life, therefore, scientists are most interested in rocky planets and moons about the size of Earth.

Venus has been described as our sister planet, but unfortunately she has succumbed to a runaway greenhouse effect, and her surface temperature is about 450°C (840°F), which is not good for life. Mars is rather cold, and without much atmosphere, and therefore exposed to vicious ultraviolet radiation from the Sun, but nevertheless there are faint possibilities for life on Mars (see page 168).

Uranus was discovered purely by accident. Herschel was simply searching the sky for new objects and stumbled upon it. Neptune, although it had been observed by Galileo and others without being recognized as a planet, was finally discovered by mathematics. Uranus was seen to have a curious and irregular orbit, which suggested that another planet was disturbing it. Using Newton's equations, John Couch Adams in England and Urbain-Jean-Joseph

'So far as hypotheses are concerned, let no one expect anything certain from astronomy, which cannot furnish it, lest he accept as the truth ideas conceived for another purpose, and depart from this study a greater fool than when he entered it'

Nicolaus Copernicus

Le Verrier in France worked out where this unknown planet should be. Adams had become interested in the problem while still an undergraduate at Cambridge, and finally produced his predictions on 1 October 1845. Independently, Le Verrier published his predictions on 1 June 1846. Both men tried to persuade the most senior astronomers in their respective countries to search for the beast, but were ignored, and it was not until the late summer of 1846 that anyone looked seriously.

British astronomer James Challis saw Neptune on 4 August, but did not realize what it was. On 23 September the German astronomer Johann Galle received Le Verrier's predictions, turned his telescope in the right direction, and found Neptune within half an hour. Who got the credit? It seems to have been shared between Adams, Le Verrier, Challis and Galle.

MOONS

And then there are the moons that orbit the planets. Mercury and Venus have no moons. We have one big Moon, only 385,000 km (240,000 miles) away, but receding from Earth at the rate of 4 cm (1.5 inches) per year. Mars has two tiny moons called Phobos and Deimos, which look for all the world like lumps of rock, or very old potatoes.

Jupiter, however, with its great mass, has attracted a huge collection of moons. The four biggest – Io, Europa, Ganymede and Callisto, are big enough to be visible through a pair of binoculars or a small telescope. They look like stars in a straight line either side of the planet, although often they are not all visible at the same time. Galileo saw them through his telescope in January 1610, and for him this was proof of Copernicus's assertion that Earth is not at the centre of the universe, for here were some heavenly bodies revolving around something other than Earth. Jupiter is now known to have at least 63 moons, and more will probably be spotted in the future. Curiously, most of these moons are rocky, and the four largest are similar in size to our Moon. Saturn, slightly less massive, has at least 56 moons, Uranus has 27, and Neptune has 13.

Our own Moon is rocky, dry and lifeless, as the American astronauts confirmed when they visited it between 1969 and 1972. It probably formed soon after Earth (4.5 billion years ago) when an object the size of Mars crashed into Earth and knocked off a large chunk of the crust and mantle. This is supported by the fact that the chemistry of the Moon's surface is uncannily similar to that of Earth's mantle; the astronauts brought back 400 kg (880 lb) of Moon rocks for examination. If our Moon ever had an atmosphere it long ago escaped into space, and the surface is left mercilessly exposed to radiation from the Sun and to bombardment by space rocks, which is why the Moon's face is so scarred and cratered.

The astronauts confirmed that the Moon's dust is powder-dry and lifeless, but astronomers have recently noticed something funny. Every now and then the surface seems to emit a luminous glow, which sometimes lasts for only a few seconds but at other times goes on for hours. Are these Lunar Transient Phenomena (or LTPs) real? And are they signs of some sort of life? Life seems most unlikely, but there are two rival theories to explain LTPs: they could be caused by gases escaping from below the surface – the Moon gasping for life – or they could be the visible result of impacts by meteorites. There are still questions to be answered about the Moon. Could there be frozen water hidden in craters? And what lies below the dusty surface?

Mars's moons are small bare chunks of rock, possibly captured asteroids, but the moons of the gas giants are quite different. The largest ones formed at the same time as their parent planets and are rocky and large – several are larger than our Moon – and several have water. These may be the most likely places for life in the solar system.

Europa is nearly as big as our Moon, and seems to have a metallic core, a rocky mantle and an outer covering of water ice, which may be many kilometres thick. This ice is highly reflective, and makes Europa one of the brightest objects in the solar system. Photographs from Galileo show that the ice is criss-crossed with cracks, some many kilometres wide, which suggest that there is either heat or movement from below. This heat might come from volcanoes on the bed of an underground ocean, from radioactive decay or from tidal forces that result from the proximity to Jupiter.

All the moons of the outer planets are cold on the surface, but they could well be warm deep down, either in the water or in the rocks – which is where primitive life may exist. We already know how life-forms known as extremophiles inhabit the most extraordinary niches on Earth – see pages 166–7). Even though Europa is the number one candidate, there is a severe problem with investigation, because of the thick layer of ice on the surface. Making a probe that can drill several kilometres down into solid ice won't be easy.

Io, one of Jupiter's four large Galilean moons, is here seen against the backdrop of the giant planet. The two appear close but are in fact three times the diameter of Jupiter apart.

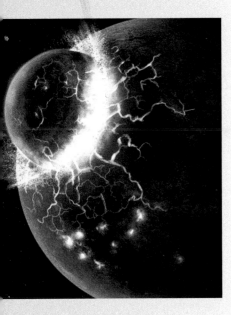

When an object the size of Mars smashed into Earth, molten material from the two bodies splashed out into space, forming a dense, doughnut-shaped ring around Earth. Eventually the debris clumped together in orbit to form the Moon.

DATA FILE: BIG MOONS OF THE SOLAR SYSTEM, COMPARED WITH OUR MOON

Planet	Moon	Distance (000 km)	Diameter (km)	Month (Earth days)
Earth	Moon	384	3476	27
Jupiter	Io	422	3600	1.8
	Europa	671	3100	3.6
	Ganymede	1071	5300	7
	Callisto	1883	4800	17
Saturn	Titan	1222	5150	16
Neptune	Triton	355	2700	6

Huygens is parked on a level piece of ground surrounded by pebbles 5–10 cm (2–4 inches) in diameter. Titan's crust is hard and thin with evidence of softer material beneath.

What news did Huygens send back from its trip to Titan? We already knew that Titan is the only moon in the solar system with a substantial atmosphere, 98 per cent nitrogen but with some methane, and traces of hydrogen, hydrogen cyanide, ethane, argon and other organic compounds. Under the influence of sunlight, these gases can react together to form more complex compounds, some of which include nitrogen. These are prebiotic compounds – the sort of chemicals that might ultimately lead to life. The atmosphere and conditions on Titan are in some ways similar to those on the early Earth, and deep-frozen information from Titan may well tell us something about the origin of life on Earth.

Titan's river beds look as though they have been scoured out by liquid, but since the temperature on the surface is about -180°C, (-290°F) this cannot be liquid water. The three simplest hydrocarbons are methane, ethane and propane. All three melt below -180°C (-290°F) and they boil at -162°C (-250°F), -89°C (-128°F) and -42°C (-43°F) respectively; so the liquid might be any of them, although methane might evaporate quickly in the sunshine. Ethane is the most likely candidate for the liquid on the surface.

The atmosphere above about 30 km (18 miles) is foggy, probably with methane droplets, and there is evidence of complex weather patterns. There is at least one substantial mountain range, about 150 km (90 miles) long; the mountain tops are covered with what looks like snow – perhaps methane snow. There appear to be lakes or seas of liquid hydrocarbons near Titan's south pole. Even though the moon must have been hit by many chunks of space rock, there seem to be few craters, which suggests that the terrain is constantly being resurfaced, possibly by water, ammonia or hydrocarbons from hot springs.

And finally, the mother ship Cassini has sent back extraordinary photographs of one of Saturn's inner satellites, Enceladus, which is a tiny moon with a diameter of only 500 km (300 miles). Dr Michelle Doherty at Imperial College in London is responsible for an instrument on Cassini that measures magnetic field, and she noticed that Saturn's magnetic field did not just pass straight through Enceladus but bent around the moon. So she persuaded her colleagues to alter Cassini's course so that it flew only 170 km (105 miles) above Enceladus's south pole. From there the photographs show plumes of water vapour escaping from cracks in the surface. The surface is smooth, very bright, and cold – about -200°C (-330°F) – so it may be water ice, and the vapour seems to be spouting out under pressure, like an array of geysers.

PLANETS OUTSIDE THE SOLAR SYSTEM

The planets in our solar system are easy to spot. Mercury, Venus, Mars and Jupiter are visible to the naked eye – although Mercury can sometimes be tricky because it is always so close to the Sun – while the outer planets can be seen with a small telescope. Other solar systems are much further away, however.

The nearest star is more than four light years away – that is about 40,000 billion km (25,000 billion miles), or nearly 300,000 times as far from Earth as our Sun. Stars shine brightly, but planets don't – they are visible only in reflected light from their suns. This means they are at least a million times fainter than their suns, and the glare from the sun makes them all but impossible to see at such great distances; indeed they are almost invisible even to the most powerful telescopes. And yet, in spite of this invisibility, more than 200 extra-solar planets, or 'exoplanets', have been found. How is it done? By cunning and skill. This is a fine example of how rapidly astronomy has advanced in the past few decades. For at least 150 years astronomers have speculated about exoplanets. The first was spotted in 1992 and now at least 20 are found every year.

The deep fracture lines that criss-cross the surface of Enceladus may be evidence of resurfacing caused by tectonic activity below. This tiny moon has a variety of landscapes, with the older surfaces pocked by impact craters.

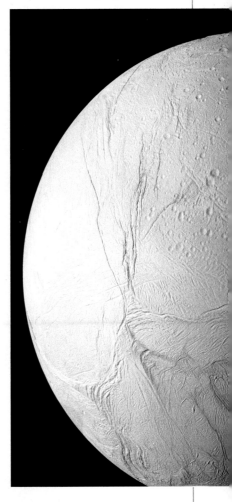

This artist's conception depicts the pulsar planet system discovered in 1992. Pulsars spin and pulse with radiation. Here, the pulsar's twisted magnetic fields are highlighted by the blue glow. Radiation from charged pulsar particles would probably stream down on the planets, causing their night skies to light up with auroras similar to our northern lights.

EXOPLANETS

The first exoplanets were found orbiting a dying pulsar with the exotic name of PSR B1257+12 in the constellation Virgo. Polish astronomer Aleksander Wolszczan discovered the pulsar in 1990, using the Arecibo telescope (see page 175). It's 980 light years from Earth, and rotates every 6.22 milliseconds – amazingly fast. Two years later, working with Canadian astronomer Dale Frail, Wolszczan discovered that radio pulses from the pulsar sometimes arrived earlier or later than expected. These anomalies in the pulsation period were due to the pull of three Earth-sized planets – possibly the rocky cores of former gas giants. These planets around the pulsar are most unlikely to harbour any sort of life, since the radiation would be lethal. It would kill anything as close as those planets.

DATA FILE: FIRST EXOPLANETS COMPARED WITH EARTH

	Pulsar distance	Diameter (km)	Mass	Length of day	Length of year
A	0.2 AU	?	0.02	?	25 days
B	0.36 AU	?	4.3	?	67 days
C	0.46 AU	?	3.9	?	98 days

This artwork shows a triple sunset on a conjectural moon orbiting exoplanet HD 188753 Ab, found in 2005. The planet orbits a triple-star system, a finding that challenges current theories of planet formation.

In 2006 the Observatory of Geneva discovered three Neptune-sized planets orbiting star HD 69830 (above), which is much like our Sun. Their existence was deduced from the gravitational wobbling of the star.

DATA FILE: FINDING EXOPLANETS

There are seven methods of detecting exoplanets:
- pulsar timing – as with PSR B1257+12 (left);
- gravitational wobble – the push-me-pull-you effect of the planet on its sun, detected by spectroscopy;
- transit dimming – the slight reduction in brightness as a planet passes in front of its sun;
- astrometry – seeing a star wobble as a result of its planet;
- gravitational lensing – the magnification by the huge gravitational effect of a star or galaxy;
- the disturbance in discs of dust, which often gather round stars;
- direct observation.

GRAVITATIONAL WOBBLE

When a massive planet such as Jupiter orbits the Sun it exerts a gravitational pull on the Sun, which by Newton's third law (action and reaction are equal and opposite) must be the same as the pull needed to keep Jupiter in orbit. This means that Jupiter does not revolve around a stationary Sun; rather the pair of them actually revolve around a common centre of gravity. Because the Sun is a thousand times more massive than Jupiter, this centre of gravity is actually below the surface of the Sun, but it is still well away from the Sun's core, and as Jupiter swings round to the 'east' the Sun will be swinging 'west' to balance the pair.

The first exoplanet to be discovered by using this phenomenon, on 6 October 1995, orbits a star called 51 Pegasi, and is called 51 Pegasi b. (The first planet near a star is now called by the name of the star followed by b – then c, d, and so on.) It was spotted by Swiss astronomer Michel Mayor and his student Didier Queloz at the Observatoire de Haute-Provence near Marseille. The local village is now called St Michel l'Observatoire. Mayor and Didier were looking for unusual features of nearby stars, using an old telescope and a highly stable spectrophotometer of their own design. They noticed that the spectral lines of '51 Peg' were moving to and fro; that is, the frequency was repeatedly getting slightly higher and slightly lower.

This looked like the Doppler effect, which causes frequencies to increase when the source is moving towards you and to decrease when the source is moving away (see page 16). This was just what the spectrum of 51 Peg was doing.

They guessed that a massive planet was orbiting 51 Peg, and causing the star to wobble to and fro, which would account for their observations. The problem was that the oscillation took place over only 4.2 days, which meant that the planet, which appeared to be almost as big as Jupiter, must be circling its sun enormously fast, and must therefore be enormously close and enormously hot – perhaps 1200°C (2200°F) on the surface. Jupiter is 780 million km (480 million miles) from the Sun and takes 12 years to make one orbit. This new planet,

This is an artist's impression of 51 Pegasi b. The intense heat of the planet may create a faint, comet-like tail pointing away from the planet's star.

even though it has roughly the same mass as Jupiter, is circling in just a few days, and must therefore be only about 16 million km (10 million miles) away from its sun. For this reason no one believed the result to start with, and Mayor and Queloz were ridiculed, until their results were duplicated by an American team.

After that many astronomers started looking for wobbling planets, and Mayor, Queloz and their American colleagues have found dozens more. This method is likely to find large planets close to their suns – now known as 'hot jupiters' – but unlikely to find Earth-like planets at Earth-like distances, even with more sensitive equipment, because their 'wobble' effect on their suns is so small as to be difficult to observe. Nevertheless it has become an established method of hunting for exoplanets. Furthermore, the technology is improving, and we can now detect Doppler shifts caused by star velocity variations of only 2–3 metres (6–10 feet) per second – walking speed.

Today, more than 20 multi-planet systems have been discovered, which suggests that there are many out there – and this again increases the probability that one of those planets might be like Earth – and might harbour life.

'We are getting much closer to seeing solar systems like our own'

Didier Queloz

Michel Mayor (right) stands with Didier Queloz before the telescope at L'Observatoire de Haute-Provence. Since the discovery of 51 Pegasi b, Mayor and his team have been involved in the discovery of many more exoplanets. In 2004 Mayor was awarded the Albert Einstein Medal for his outstanding contribution to astrophysics.

TRANSIT DIMMING

The next most important hunting technique in finding exoplanets is to look for transits. If a planet passes between you and a star this is called a transit. The first ever observed was a transit of Venus across the face of the Sun on 24 November 1639, when a young English curate named Jeremiah Horrocks saw a little black dot moving across the Sun. He observed it for half an hour before the Sun set.

The same method is used to find exoplanets. A little of the starlight is blocked by the planet, so the star appears to dim slightly. The effect is small – unless the planet is enormous – but even a 1 per cent dimming can be observed, and the observation can be made repeatedly to check. Today, light curves are used to detect dimming. This diagram shows what the light curve looks like. The frequency of the dimming tells you how long the planet takes to go around the star, and therefore its speed. The time taken to get fully in front of the star (B) tells you the diameter of the planet, and the total transit time (A) tells you the diameter of the star. Amazing what can be deduced from a single curve.

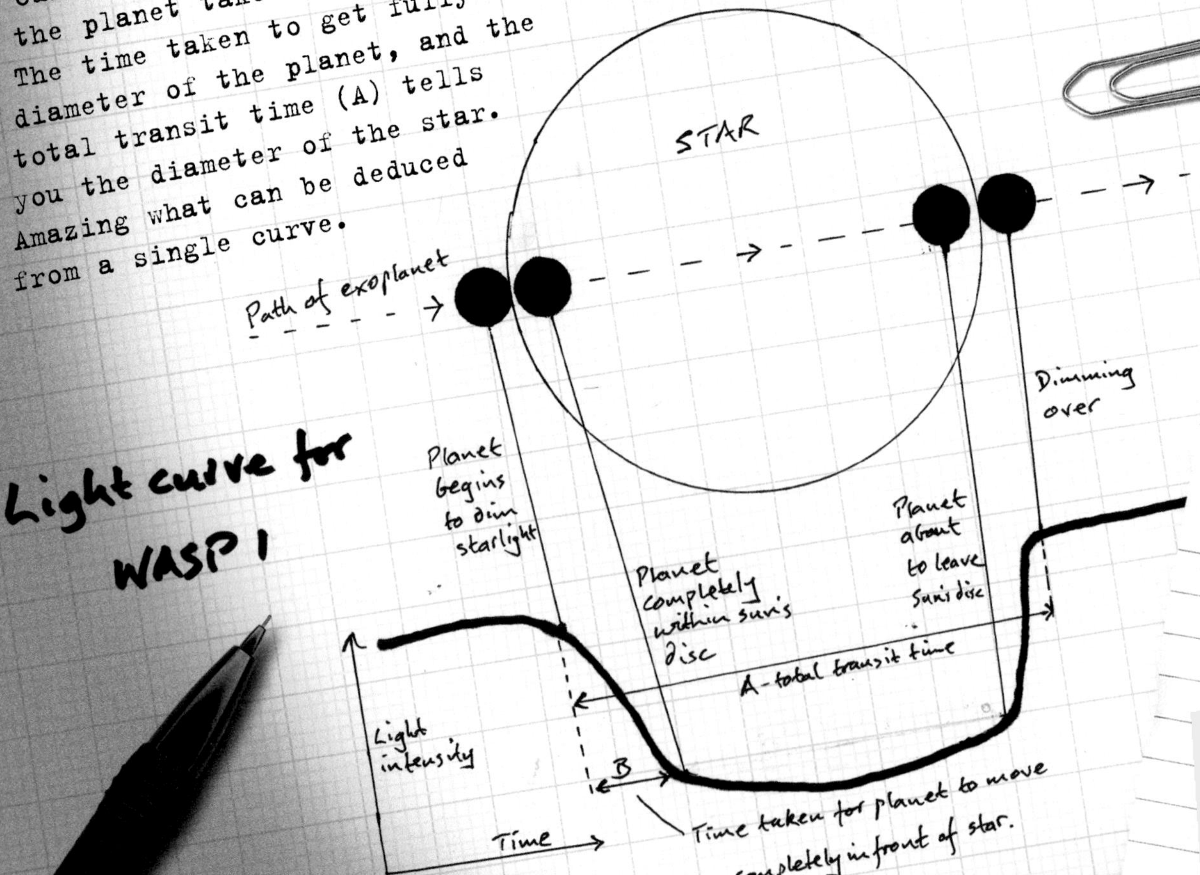

STAR

Path of exoplanet →

Dimming over

Light curve for WASP 1

Planet begins to dim starlight

Planet completely within sun's disc

Planet about to leave sun's disc

A - total transit time

Light intensity

B

Time →

Time taken for planet to move completely in front of star.

In our demonstration an 'exoplanet' is about to transit its star. We will be able to tell something about the nature of the planet from the degree of dimming as it passes the star.

Adam monitors the star's brightness with a light meter. Any variation in brightness and he will know there is an object passing between him and the star.

Transits of Venus happen about twice a century. The last one was in 2004; the next will be in 2012; after that not until 2117 and 2125. Meanwhile Mercury transited the Sun on 8 November 2006 and will do so again in May 2016.

This transit of Venus was snapped on 8 June 2004.

As the star dims, Adam's light meter records the drop in brightness. From how long it takes to drop he can calculate the planet's diameter. He times the dimming and from this can calculate the diameter of the star.

115

SuperWASP

We were astonished to find not a huge telescope in a vast building but a bunch of eight digital cameras in a garage with a sliding roof.

Astronomy is both work and hobby for Don Pollacco. As he says, other planets were science fiction when he started; now they're for real.

SuperWASP's neighbour, the giant William Herschel Telescope, is the largest in Western Europe, and verifies SuperWASP's findings.

The principal performer in the transit detection method is SuperWASP, which comprises one telescope on the rim of an old volcano at La Palma in the Canary Islands, and another at Sutherland Observatory in South Africa. WASP stands for Wide Angle Search for Planets, and the instrument was designed and built by Don Pollacco and his colleagues with a tiny budget. 'SuperWASP' seems to us a splendid name for such a piece of ingenuity. Its great power lies not only in its sensitivity but also in the fact that it can survey vast numbers of stars at the same time.

The telescope's cameras are mounted in a 4×2 rectangle on a robotic arm looking at adjacent bits of sky, so that between them they cover an immense field of view. They each take one 10-second exposure and one 30-second exposure (to pick up fainter stars), and then the robot arm moves them on to another patch of sky. The telescopes sweep the entire sky several times every night, and every time they pause, each camera photographs 50,000 stars. All the data are sent back over the Internet to computers in the UK, and by measuring how the brightness of each star varies during the night the software can pick out possible candidates for stars with transiting planets. For each star the computer has a string of brightness values, which it can plot as a light curve – a graph of brightness against time. It is looking for is a tiny dip in the brightness that is repeated at regular intervals.

During its first full year, SuperWASP scanned 6.7 million stars, and found 18,000 stars with apparent dimming. After looking critically at the data, Don and his team narrowed this down to 100 strong candidates for stars with planets. Two of those have so far been examined using bigger telescopes, and found to be genuine; the planets are called WASP1 and WASP2. Don is confident that

A short distance down the mountain from SuperWASP is the 4.2-metre (14-foot) William Herschel Telescope, which checks out some of the possible planets discovered by its precocious young cousin. By zeroing in on a candidate, the WHT can examine its brightness with precision, and the shape of the light curve gives much information about the new planet (see page 114). At the same time the WHT can measure the spectrum of the star, and may as it transits be able to measure the spectrum of the planet too, and discover whether it has an atmosphere, and if so what its composition is. When Don or one of his colleagues gets the chance to look at one of the SuperWASP possibles with a big telescope they do something rather cunning. Once they have locked onto the right star, they deliberately defocus the telescope a little. The star is a bright point source of light, so it hits only one pixel of the detector, which is all too easy to saturate; so measuring a small change in brightness is difficult. When the telescope is defocused, the image of the star spreads out into a disc; so it falls on many pixels, and the detector can receive many more photons before being overloaded. Therefore small changes in brightness become much easier to measure.

This is SuperWASP's array of cameras. Each has a 200-mm (7.9-inch) f/1.8 lens 11 cm (4.3 inches) in diameter.

DATA FILE:
KEPLER MISSION

A 95-cm (37-inch) diameter space telescope, Kepler will trail Earth in a solar orbit. This will provide the maximum of stability and the minimum of interference from radiation and gravity. Kepler will look out of the plane of the ecliptic, so that the Earth, Moon and Sun do not get in the way, and will, like SuperWASP, repeatedly photograph thousands of stars, in the constellations Cygnus and Lyra, in the search for exoplanets.

'The strength of this [transit dimming] technique is that there is very little physics in it; it's mostly geometry and geometry is quite well understood'

Don Pollacco
SuperWASP astronomer

There are three problems with SuperWASP. The first is that, like all optical telescopes, it can see the stars only at night. A planet might transit its star during the day, and SuperWASP would never know. Indeed, a planet could possibly transit its star every single day at noon. What astronomers would really like is a telescope at the South Pole, which could observe continuously for six months at a time, plus another in the frozen north – Norway, Russia, Canada or Alaska – to cover the other half of the year.

The second problem is that although it is clearly good at finding hot jupiters, it will probably never be good at finding 'cold jupiters' or Earth-like planets. In our solar system Jupiter takes 12 years to orbit the Sun, and the chance of an observer from outside the solar system observing for 12 years and spotting the dimming is small indeed. Furthermore, any observer not in the plane of the ecliptic (the plane of Earth's orbit around the Sun) would see no dimming at all. Even to see Earth you might have to observe for a whole year, from within the ecliptic plane, and then the dimming would only be about 0.01 per cent – too small to be detectable.

The third major problem is the twinkling of stars caused by turbulence in the atmosphere, which is a problem for all Earth-based astronomy. Being certain of a dimming of less than 1 per cent when the stars are twinkling like Christmas lights is a tricky business. It is a formidable computing job to compare the brightness of all the stars on a given photograph so as to be sure of the dimming of one star relative to thousands of others, and astronomers take multiple photographs to improve the accuracy of this process. To spot Earth-sized planets they need to be able to observe a dimming of only 0.01 per cent – that's 100 times less than they can see now. This might just be possible if they could use a telescope in space, free from Earth's atmosphere.

That is the object of the French-backed Corot mission, launched in the last days of 2006 – a 30-cm (12-inch) telescope designed to search for rocky planets from Earth orbit – and also NASA's Kepler probe (see box), to be launched in 2008 on a four-year mission.

BRILLIANT DEDUCTIONS

The transit method can do much more than simply detect planets: it can provide information about the inclination of the planet's orbit, its mass and density, and even something about the weather systems and the temperature.

From minor variations in the transit time of a hot jupiter – the time it takes to cross the face of its sun – the SuperWASP team may be able to deduce the presence of smaller planets in larger orbits, whose gravity could affect the orbital speed of the giant. This was in effect how Neptune was found – from variations in the orbit of Uranus. The team also hope to learn general information about the formation of solar systems – how the sizes and masses of planets vary, how big planets get so close to their suns, and what happens to them – do they get swallowed up, or do they evaporate into space? So far all the solar systems they have discovered contain hot jupiters; so they are quite unlike our own solar system. As a result no one yet knows whether our type of system is typical, or unique.

Of all the 200 new planets that have been discovered, some stand out as being particularly interesting. The red dwarf star Gliese 876 is known to have at least three planets, one of which is only 7.5 times as massive as the Earth. Exoplanet HD 209458b was discovered by gravitational wobble in 1999, and was then spotted again by transit dimming, which was strong evidence that the gravitational wobble method really does detect planets. Two years later, observations from the Hubble Space Telescope showed that this planet has an atmosphere that contains the element sodium.

The Observatoire de Haute-Provence has acquired a powerful new instrument called SOPHIE, designed to detect gravitational wobbles. The partnership of SOPHIE and SuperWASP is especially valuable, since SuperWASP finds possible planets and measures their size, while SOPHIE confirms that they are planets and measures their masses. In its first four nights of operation, SOPHIE found SuperWASP's first two planets.

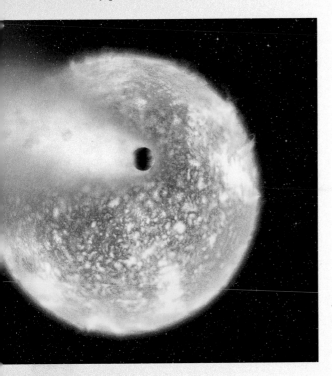

An artwork of exoplanet HD 209458b shows it transiting its star. The planet is 150 light years from Earth in the constellation Pegasus.

This artist's impression of OGLE-2005-BLG-390Lb shows the exoplanet orbiting its red dwarf star. The surface of the exoplanet is a chilly -220°C (-364°F).

GRAVITATIONAL LENSING

In his famous paper on general relativity, Albert Einstein said that gravity does not really make two masses attract one another; instead it distorts space in such a way that the two are pulled together. One consequence of this is that a massive star distorts the space around it rather as a lens distorts the light going through it. Indeed the space around the star actually acts as a lens, attracting the light rays towards its centre, and thus magnifying the image of a distant object on the far side. This happens when by chance a star passes in front of a distant planet, which flashes bright for between a few minutes and an hour or so.

Using this method, a planet called OGLE-2005-BLG-390Lb was discovered on 25 January 2005; it orbits a red dwarf 21,500 light years away, near the centre of our galaxy. It seems to have a mass only 5.5 times that of Earth, and is nearly three times as far from its star as we are from the Sun.

OTHER METHODS

Direct observation is exceedingly difficult; only the most powerful telescopes can hope to see these dim and distant objects, but in 2005 observers at the VLT in Chile managed to photograph planet 2M1207A in orbit around a brown dwarf. This was the first image ever of a planet in another solar system.

Most of the other detection methods are in their infancy. All are most effective at finding massive planets close to their parent stars, and most of the planets discovered so far have been larger than Jupiter and much closer to their stars than we are to the Sun. However, the fact that they exist at all has opened up an entire new area of astronomy; meanwhile the technology is improving quickly, and no doubt some Earth-like planets will be found in due course.

IMPOSTORS

Astronomers searching for exoplanets are occasionally led astray by impostors, especially brown dwarfs. Fifty years ago the classification was simpler: stars were things that shone; objects orbiting stars were planets; objects orbiting planets were moons. However, things have become more complicated, which is typical of scientific endeavour. The more questions you ask, the more questions you end up with, and the more complicated everything becomes. We now have a great range of stars, including neutron stars, pulsars and black holes. There are also stars called brown dwarfs, which are big balls of stuff that don't quite have enough mass to light their own nuclear fires (see box).

Another class of impostor is the variable star. Variable stars were known to the ancients – Algol, in the constellation Perseus, was given its name ('the ghoul') by Arab astronomers, because even with the naked eye it can be seen to wink; every three days it appears to dim, and then brighten again. In May 1783 amateur British astronomer John Goodricke suggested to the Royal Society that this

The red blob in this historic infrared photograph is the first direct shot of an exoplanet. The blue blob is the brown dwarf star it orbits.

DATA FILE: BROWN DWARFS

Brown dwarfs are balls of hydrogen that do not quite have enough mass for the process of nuclear fusion to begin; so as stars they do not shine – at least not with visible light. They do shine in the infrared (above), but that is hard to detect from Earth, because the radiation is absorbed by our atmosphere. Curiously, all brown dwarfs seem to be about the same size as Jupiter, but they have masses perhaps 50 times higher than Jupiter's; so if one is orbiting a bright star it may be detected by gravitational wobble and mistaken for a planet.

BIOG FILE: JOHN GOODRICKE

John Goodricke, born in the Netherlands in 1764, spent most of his life in England, and stayed with his parents in York. A bout of scarlet fever in his youth left him profoundly deaf for the rest of his life. He was a skilled astronomer, and his records were so precise that historians have been able to deduce that he observed looking south from the eastern window on the second floor of The Treasurer's House, a fine medieval building near York Minster. At the age of 21 in 1786 Goodricke was elected a fellow of the Royal Society, but he died before the news reached him. He had spent too much time observing on long cold nights, and succumbed to pneumonia.

happened because a dark body was passing in front of the star, and this contribution won him the Society's Copley Medal in that year. We now know that Algol is a binary pair – two stars revolving around one another. One is brighter than the other, and they happen to revolve in a plane that includes the line of sight from Earth, so that once during each revolution the dim star blocks our view of the bright one, which is why Algol appears to blink. In fact there is a third star much further away revolving around the pair, but that does not affect the winking.

SuperWASP would not be able to distinguish between an orbiting planet and a variable star, but SOPHIE the gravitational wobble detector can be brought to bear, and can measure the mass of the object, which decides at once between a star and a planet. This is what makes the combination of SuperWASP and SOPHIE so powerful. Since between them they measure both the diameter and the mass of a planet, we can also know the density – the mass divided by the volume – and this proves fascinating. The density of the Earth is 5.5 g/cm^3, which is about halfway between the densities of aluminium and iron.

The rocky planets are much denser than the gas giants (see box opposite), and Saturn is such a lightweight that it would float in the bath – if you could find a big enough bath. The density of water is 1 g/cm^3, and anything less dense than water will float. (Guessing the densities of various fruits and vegetables is good fun. Write down first whether you think apples or oranges or bananas or tomatoes will float or sink in water, and then try them out.)

WATER

One of the major requisites for life as we know it is water, ideally in liquid form on the surface of the planet. Water may originally form when the planet itself is created, or it may arrive afterwards from space in the form of ice, since liquid water cannot exist in the vacuum of space. The planet would have to be cool enough for vaporized water to condense, which it couldn't do on a hot planet. One possibility is that all water on planets, including Earth, arrived in the form of those dirty snowballs, the comets. Comets are a mixture of ice, rock and dust or soot that condensed into lumps outside the solar system in regions known as the Kuiper Belt and the Oort Cloud. These are a long way from the Sun, and any steam around would readily condense on a passing comet.

Comets orbit the Sun in highly eccentric orbits; one end of the orbit may be way out beyond Neptune and the other as close to the Sun as Mercury. These orbits are usually long – Halley's Comet returns every 75 years, and has been observed in our skies since 240 BC. As a comet approaches the Sun, the growing heat begins to melt some of the ice, and the resulting debris of ice particles and dust is pushed out away from the comet by the solar wind, forming the comet's 'tail', which always points directly away from the Sun.

DATA FILE: PLANET DENSITIES

	Density (g/cm³)
Mercury	5.4
Venus	5.2
Earth	5.5
Mars	3.9
Jupiter	1.3
Saturn	0.7
Uranus	1.3
Neptune	1.6

This 1986 image clearly shows the whitish dust tail of Halley's Comet – next to be seen in 2061.

BIOG FILE:
EDMOND HALLEY

Edmond Halley was born in 1656. In 1676 he left Oxford for the island of St Helena in order to survey the stars of the southern sky. Halley became one of Isaac Newton's closest friends, and persuaded him to write the important book *Principia*. In 1682 Halley observed a comet with an unusual retrograde orbit – it went round the Sun the 'wrong way'. He realized that this comet had been seen before, in 1456, 1531 and 1607. He correctly predicted that it would return in 1758, and it has been called Halley's Comet ever since, although the Chinese had observed it in 240 BC. Halley was made Astronomer Royal in 1720 and died in 1742.

This illustration shows meteorites and comets striking the young Earth during its formation and bringing water ice to the surface. The Moon appears much larger in the sky as it was nearer to Earth at this time.

The young Earth was hit by large numbers of meteorites and comets, which kept it hot – perhaps even molten on the surface – but which also brought vast amounts of water. Some of the water boiled off into the atmosphere. This later condensed and fell as rain to form the oceans.

Water is difficult to detect on other planets, but in our solar system we know there is no water on Mercury or Venus; there is some on Mars, mainly below the surface, and none at all on the Moon, except perhaps at the poles. The gas giants have no water, but some of their moons have large frozen oceans, notably Jupiter's moons Europa, Ganymede and Callisto, and Saturn's moon Enceladus.

ATMOSPHERES

The first indication that other planets might have atmospheres came in the seventeenth century, when astronomers noticed that when it transited across the Sun, Venus seemed to have a fuzzy edge.

In the 1860s English astronomer Norman Lockyer recorded a spectrum of the thin strip of sunlight that was left peeking around the edge of the Moon during a solar eclipse, and noticed there was a bright spectral line from some unknown element.

He guessed that it was a metal, and called it helium, from the Greek work for Sun, *helios*. Later the element was discovered on Earth by the Scottish chemist William Ramsay and was found to be a gas. It is the second commonest element in the universe – after hydrogen – and is formed in stars by the fusion of hydrogen atoms and on Earth by radioactive decay.

Lockyer's discovery prompted scientists to look further at the spectra of atmospheres of stars and planets, but for planets this is not simple because they do not shine like stars.

Amazingly, some of the constituents of Earth's atmosphere can be deduced by examining the spectrum of the dark part of the Moon. When the Moon is just a slim crescent, you can often see the rest of it as a sort of ghostly shadow; that is because it is lit by 'earthshine' – light reflected from Earth. Examination of the spectrum of earthshine shows spectral lines of oxygen and carbon dioxide.

For exoplanets the task is still more challenging, since they are so far away that they cannot even be observed directly. However, when a planet transits its sun, the sunlight will shine through the atmosphere at the edge of the planet, and this is enough to get a spectrum. The instruments are so sensitive that the observers can tell a good deal about a planet's atmosphere from a single pixel of light.

One feature scientists look for is the 'red edge' characteristic of Earth's green plants. Green plants absorb most of the visible light that falls on them, and use the energy from the light to drive the process of photosynthesis. However, this absorption cuts off sharply just beyond the visible spectrum, in the near infrared, probably because if the plants absorbed strongly in the infrared there would be a danger of overheating. So just outside the visible spectrum, at a wavelength of about 700 nm, the amount of light reflected from greenery rises sharply from about 5 per cent to 50 per cent, and this abrupt change is called the red edge. For this reason grass and other leaves appear bright in infrared photography. A red edge in the spectrum of an exoplanet would be a strong suggestion that there might be living plants there.

PLANETS ON THE LOOSE

We think of the natural habitat of planets as being in orbit around a sun, but occasionally – perhaps as the system is first forming – it seems that a planet gets flicked away from its sun by gravitational interaction with another planet, or by a star that passes too close. Then it is likely to wander alone through the cold dark of space, for the chances of being picked up by another sun are small. There is some argument about whether these objects should be called planets, since planets are supposed to orbit stars; so purists call them 'interstellar planetary-mass objects'.

Some scientists specialize in hunting for these rogue planets, and a few even believe that there may be more rogue planets than planets in solar systems. Unfortunately rogue planets are extremely hard to find because they are completely dark – not even illuminated by a sun – and we have no idea where to look.

However, there is a possibility that such a planet might harbour primitive life. Although it would not be heated by a sun, enough heat might be generated by geothermal activity and radioactive decay to keep the surface above 0°C (32°F), so that liquid water could exist on the surface. Furthermore, without ultraviolet radiation from a sun, an Earth-sized rogue planet could easily retain an atmosphere, mainly of hydrogen and helium.

VIRTUAL PLANETS

In a leafy suburb of Los Angeles a team of NASA scientists is making its own planets inside computers. This is the Virtual Planetary Laboratory, or VPL. Since 2001 this group of astronomers, chemists, spectroscopists, climatologists and other scientists has been inventing a range of imaginary Earth-size planets with various ages, temperatures, atmospheres and so on, in the hope of making it easier for astronomers to spot planets that are most likely to harbour life.

They also hope that as more and more information about real exoplanets is collected, they at VPL will be able to match this information with their theoretical models, and tell the astronomers more about the planets they have found.

TECHNOLOGY EQUALS KNOWLEDGE

Fifteen years ago we knew of only the seven other planets in our solar system, and neither they, nor the smaller dwarf planets, nor their moons, looked at all promising as potential homes for life. All that has changed dramatically. Several planets and some moons look like possible habitats for primitive life, and what is more we have found more than 200 other planets in different solar systems.

In other words, improving technology – and skill – has vastly increased our knowledge of other worlds around us, and has surely brought closer the time when we discover some form of life on one of them.

An artist has depicted a rogue planet faintly illuminated by the glow of a nebula but not by a sun. Dark, rogue planets moving independently through space have often featured in science fiction but have as yet eluded astronomical observation.

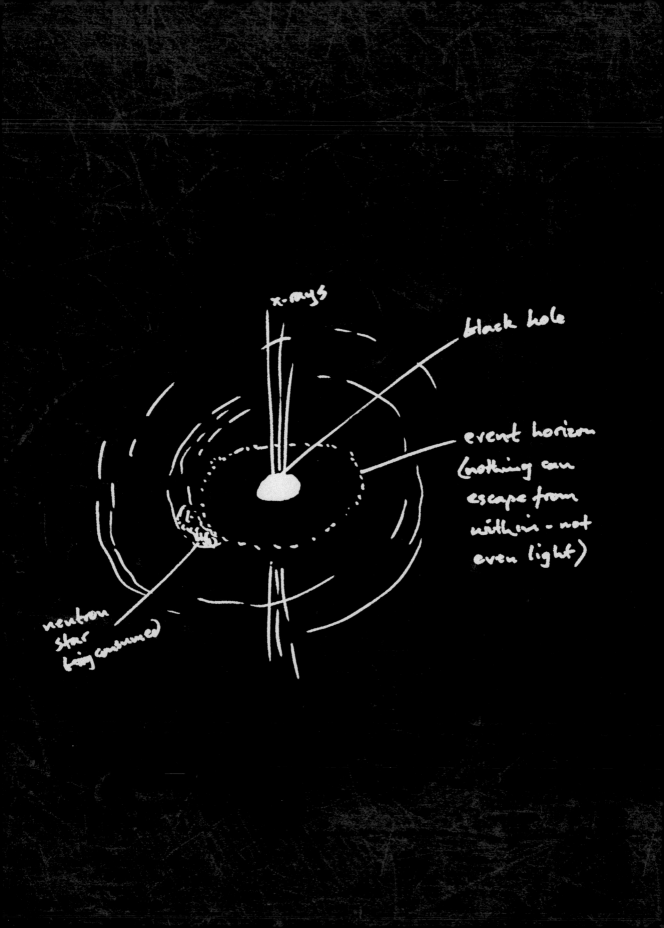

5 / VIOLENT UNIVERSE

If you could go back a few generations and ask your great-great grandmother how the universe appeared to her, her report would contain few surprises. She would see the same constellations of stars that you see, the planets doing much the same things, and the Sun and Moon rising and setting as they always do. Go back even 5000 years, and you would find little difference. Other than the odd comet, meteor shower or eclipse of the Sun, the universe as seen by Earth-dwellers appears unchanging and predictable.

But the notion of a steady and unchanging universe is a complete illusion. In this chapter you'll see that space is filled with rapid, violent events – from exploding stars to boiling magnetic storms on the surface of the Sun, from massive black holes gobbling up matter to distant explosions that emit more energy in a few seconds than our Sun will put out in a lifetime. And in case you thought the violence was all a long way away, you'll read about the team practising to save Earth from asteroid impacts, something that could happen for real within 30 years from now.

DATA FILE: BIRTH OF THE CRAB NEBULA

Inscriptions in North America have helped confirm the date of the 1054 guest star. The image above shows a hand print, a crescent Moon and a star. Because there is no written record linking this with a particular event, the rock image is all we have. Two bits of evidence help to make it convincing as a supernova record. Most interesting is what the Moon is doing in the picture. We now know that on 5 July 1054, the crescent of the new Moon would have been seen close to the almost equally bright new star – but only in North America. The combination of crescent Moon and star in daylight only happened once, and would have been visible from near Chaco Canyon. This image hasn't been dated, but pottery from nearby, also containing a record of the 'star', has been. It was in a style only used by Anasazi Indians around the time of the supernova; and from carbon dating we know that it was made between 1050 and 1070.

THE GUEST STARS

Our ancestors lacked the technology that might have provided evidence of the violent nature of the universe. But the brilliance of Chinese astronomers 1000 years ago has made it possible to tell one story that has taken hundreds of years to unfold, with new twists added as new instruments for observing became available.

Although the vast majority of stars appeared fixed, Chinese astronomers noticed that occasionally a new 'guest star' would appear, then fade and disappear again. There are records of guest stars going back to about 500 BC, and, fortunately for our story, they were recorded pretty accurately. In chapter 1 we mentioned the most significant, which appeared in 1054 and was recorded by Yang Wei-Te, a Chinese court astronomer. The date of its appearance has now been ascertained as 4 July 1054, a date confirmed by other evidence from North America (see box). This particular guest star was in the constellation Taurus, but what made it special was its brightness and its duration. It was visible in daylight, even at noon, and may have been about four times as bright as Venus. In the night sky it would have been only a little dimmer than the Moon, and by far the brightest star. Then it started to fade. After 23 days, it was no longer visible in daylight. The astronomers continued to record its night-time appearance for nearly two years. Then on 17 April 1056, 653 days after its first appearance, the guest star disappeared.

Nearly 700 years later, in 1731, the English astronomer John Bevis recorded a nebula, a faint blurry patch in the sky, visible only through his telescope. The same object was spotted by the French astronomer Charles Messier in 1758 while he was on the look-out for Halley's Comet, which had been predicted to return that year. Messier eventually realized that this nebula in the constellation Taurus was the same one seen by Bevis. When Messier began compiling his catalogue of nebulae, this was the first object in it and it is still known as M1 (for 'Messier' 1). There was some speculation about what exactly the nebula was. Messier referred to it as a 'whitish light, elongated like a candle flame'. Others, including the great William Herschel, referred to it as 'nebulous', and thought that with a bigger

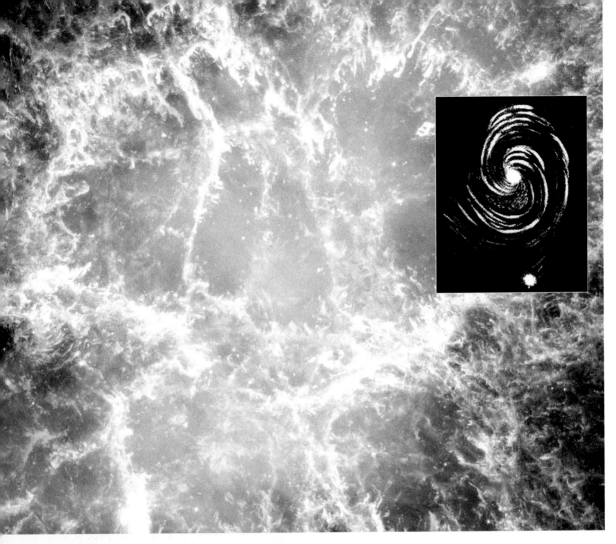

The complex structure at the heart of the Crab Nebula is revealed by modern telescopy. Inset is Lord Rosse's 1844 sketch of the nebula.

telescope it might be resolved into stars. A century later, in 1844, the aristocratic Irish astronomer Lord Rosse turned his giant telescope and considerable observing experience to M1. This was in the days before astrophotography, and Rosse was a great sketcher. His drawing showed an oval shape with 'claws', and this led to M1 being given its name, the Crab Nebula.

Later in the nineteenth century more technology was brought to bear. First, crude spectra of the

Crab showed that it is indeed made of gas, not stars. In 1892, the first photograph was taken. But although both spectra and photographs of M1 would improve year by year, the real breakthrough came in 1921 when American astronomer Carl Lampland, who was one of the first people ever to see Pluto, was working at the Lowell Observatory in Flagstaff, Arizona. Comparing photographs of the Crab taken at different times, he saw that it was growing. Between different photos, there were clear differences in size. Almost immediately this led to a question: when did the Crab start expanding? First calculations suggested that if the Crab had expanded from nothing, it would have begun about 900 years before, give or take a decade. Finally, the Swedish astronomer Knut Lundmark made the connection: the Crab was in the right place and originated at about the right time to be the guest star of 1054.

The Crab Nebula is absolutely huge. Currently about 10 light years across, it is expanding at a rate of 1800 kilometres (1100 miles) a second – and all of it is made from material that was ejected from the star that exploded before 900 years ago. A burning star is a nicely balanced system: matter is pulled inwards under by the star's own gravity, but is kept up by the pressure that comes from burning hydrogen in the star's centre. Eventually the hydrogen runs out, and the core of the star collapses in on itself. What happens next depends on how big the original star was. In the case of the Crab Nebula the star was very large, and the core would have been incredibly dense. Under this pressure, more nuclear reactions followed, creating heavier and heavier elements and a denser and denser core. Eventually iron was produced, and this was another tipping point: the core then became so dense that within perhaps one day it collapsed again, suddenly releasing a huge amount of energy. The result was a colossal explosion, ejecting most of the newly made elements out into space, and leaving an ultra-dense star in the middle. The light of this explosion, reaching China some 5000 years later, is what Yang Wei-Te and his colleagues noticed and recorded so faithfully.

Today these huge star explosions are known as supernovae. Because a star will explode in this way only when it has reached a particular condition, some types of supernova tend to behave in a very predictable way. As a result they have become a useful tool for astronomers. Edwin Hubble found the predictable brightness of Cepheid variable stars a handy reference point when he set out to measure the distances of galaxies. In the same way supernovae can also be used as reference points, and over much greater distances because they are so incredibly bright, to measure the rate of the universe's expansion. As we have already seen, recent evidence suggests that the expansion is now accelerating. Measuring the brightness of distant supernovae will confirm by just how much. The

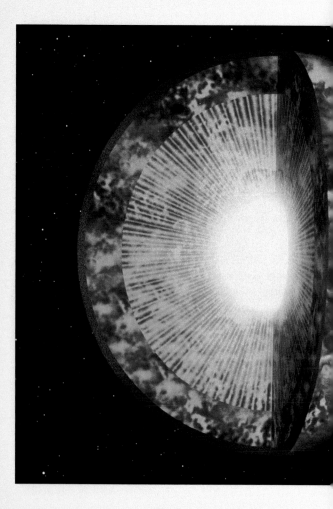

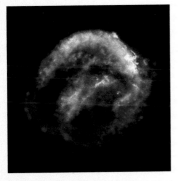

This is an X-ray image of the remnants of Kepler's supernova. It appeared in 1604 and was visible to the naked eye for 18 months.

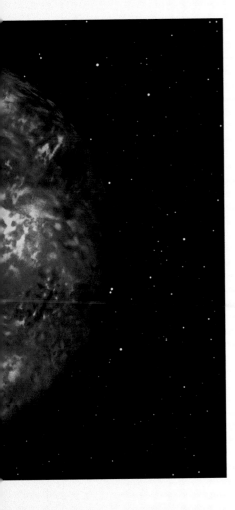

discovery that the universe's expansion is accelerating is highly significant for cosmology: first, because something has to be driving the expansion – the mysterious so-called dark energy – and second, because it has given us a clearer understanding of the universe's ultimate fate. The universe will not continue as it is, but will eventually reach a stage where it is so expanded that everything in it will be stretched out of existence. The so-called 'Big Rip' might be the final act of violence. It does not mean the end of the universe, just the end of the universe as we know it. Britain's Astronomer Royal Martin Rees, is fond of summing it up with a Woody Allen quote: 'Eternity is very long, especially towards the end.'

THE MYSTERY GAMMA RAYS

On a cosmic scale, the bangs that make spectacular nebulae such as the Crab are by no means the biggest. Curiously the really big bangs prove much harder to spot – and the discovery of the first of these happened by accident.

In the paranoid Cold-War atmosphere of the 1950s and 1960s, trust between the United States and the Soviet Union was in short supply. This was the time when both superpowers increased their arsenal of nuclear weapons to counteract any possible threat from the other. There had been about 50 tests of nuclear weapons between the end of the Second World War and 1953, and by then there was growing concern about the health consequences of nuclear test explosions in the atmosphere. Although many people favoured banning nuclear weapons altogether, it was impossible to get either side to make the first move. Eventually, in 1963, there was an agreement to ban testing of nuclear weapons in the atmosphere. When a nuclear bomb explodes, there is an immediate and dangerous flash of radiation: this affects only the local area. But the bomb also produces radioactive dust, which can be carried by the wind for thousands of kilometres. During the time nuclear tests were carried out in the

The core is the region that produces a star's energy. Next layer out is the convective zone, where energy is carried to the surface. The photosphere is the visible surface, where light is emitted.

atmosphere, the radiation they produced probably killed thousands of people worldwide.

So even though the ban on tests in the atmosphere had no effect on the number of nuclear weapons in the world, it was at least a step forward for health. But there was a problem. Since neither side trusted the other, how would they know that each was keeping their part of the bargain? The United States was concerned that the Soviet Union might cheat by carrying out tests in the atmosphere, or, worse, send nuclear weapons into space. Such was the climate of fear that some people even thought it possible that weapons could be hidden behind the Moon.

Above, the Vela 5A and 5B satellites are being prepared in a clean room. Below, a Vela satelite in orbit. Gamma ray bursts aside, a Vela satellite did in fact detect the telltale double flash of a nuclear explosion over the Indian Ocean in 1979. Details of the incident remain classified.

The American solution was to build the Vela satellites, which were sent up from 1963 to detect nuclear explosions not only on Earth, but in space. They could detect X-rays, and also gamma rays, the deadliest and most powerful radiation from nuclear explosions. Satellites were used in pairs, in part to cover the whole Earth, and in part because simultaneous readings by both satellites would confirm that it was a 'real' event. For four years there was nothing interesting. Then, on 2 July 1967, the first of a new and more sensitive generation of satellites, Vela 4a and 4b, reported a gamma-ray reading.

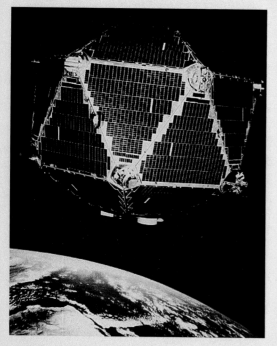

Back at the Vela headquarters at Los Alamos, New Mexico, there was immediate concern: the gamma rays had come from space. Were the Soviets really detonating nuclear weapons in orbit? Concern then turned to bafflement as it became clear that the gamma-ray reading did not follow the pattern expected from a bomb. More cases quickly followed and it was soon obvious that no one could test nuclear bombs at such a rate. The cases were classified top secret and filed away.

Six years later, in 1973, the astronomy world got a shock with the publication of a paper revealing a new and mysterious phenomenon, which would become known as the 'gamma ray burst'. The paper came from the Los Alamos scientists running the Vela programme. Finally confident that these were not Soviet weapons, they declassified 16 previously top-secret gamma ray bursts (GRBs), all of which seemed incredibly powerful, and all of which came from space. Could the world's space scientists work out what they were? A year later, the Soviets confirmed the results. They too had been worried that the United States would break the test-ban treaty, and had sent up their own gamma-ray detector satellites, which had spotted the same thing.

The GRBs were fascinating because there was no way to explain them. Most surprising was their power: clearly radiation as strong as this had to come from inside our own galaxy. Throughout the 1970s and 1980s, scientists interested in GRBs begged space on satellites and spacecraft taking part in other missions. The gamma-ray detectors they carried tried to find out where the GRBs were coming from. The work was not conclusive, though it became clear that they did not come from any known objects. Then, in 1991, a purpose-built spacecraft was launched, NASA's Compton Gamma Ray Observatory.

Although it could detect gamma rays from any direction, the purpose of this new mission was to pin down once and for all where in the Milky Way these huge bursts of energy were coming from. Already there was intense speculation about what sort of giant, dense stars might produce such immense energy when they collapsed. But what was needed was evidence. Over nine years the Compton Observatory picked up 2700 bursts. But the results did not look right. Scientists were expecting to see bursts concentrated along a line – the plane of our galaxy the Milky Way, whose disc we see edge-on. But there was

'Gamma ray bursts announce themselves not with a starlike discharge of energy in all directions, but with a telltale "torchbeam" or jet of invisible radiation'

The Guardian

The Moon seen in gamma rays by the Compton Gamma Ray Observatory. Curiously, the Sun is dimmer than the Moon at gamma-ray wavelengths.

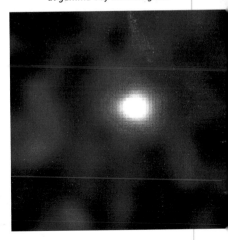

BIOG FILE:
LORD REES OF LUDLOW

Lord Rees of Ludlow, better known as Martin Rees, was born in 1942. He is one of the most accomplished astronomers on the planet – and an expert on black holes. He was not only the Plumian Professor of Astronomy at Cambridge University, then the Astronomer Royal, but is now President of the Royal Society, which makes him Britain's top scientist. Rees was the first astronomer to propose a possible solution for the origin of gamma ray bursts, suggesting that the intense flashes are emitted in a similar way to the narrow beams of radiation thrown out by black holes – like a narrow torch-beam, or a lighthouse scanning the sea. Rees has also made important contributions to the origin of the cosmic microwave background radiation and to theories concerning galaxy clusters and formation.

no concentration at all. The results appeared completely random. It was as if the GRBs were coming from all over the sky.

This result was extremely difficult for scientists to accept because it seemed theoretically impossible. If the GRBs did not come from within our galaxy, or from any nearby galaxy, then they must instead be extremely distant events, happening way out towards the edge of the universe. The most surprising thing about GRBs was their intensity, and that is why most people assumed that they had to be happening nearby. Gamma radiation, like other radiation, would get weaker as it spread out through space. But if this intense radiation came from something very distant, then the event that caused it must be impossibly violent.

Einstein's most famous equation was the problem. $E = mc^2$ was by the 1990s well tested and used to explain some of the violent events in the cosmos. The E in $E = mc^2$ is energy, the m is mass, and it is multiplied by the speed of light, c, squared – a huge number. Einstein's brilliant but mind-boggling achievement was to put an equals sign between them. In everyday life, energy and mass don't seem to be in any way equal.

Einstein was saying that in the right circumstances energy can be turned into mass, and mass can be turned into energy. To Einstein's great regret, this was the inspiration for nuclear weapons, where atoms are split and destructive energy is released. In the largest physics experiments, a huge amount of energy is turned into particles that weren't there before (see pages 20–5). And when big stars collapse, some of their mass is turned into the energy of a supernova. But when the sums were done, the energy of a GRB was simply too big. There was no conceivable object whose collapse could spew out enough energy to produce the intense GRBs witnessed from Earth. The scientists were completely stumped.

The scientific knight in shining armour who rescued the distressed GRB researchers was the Astronomer Royal himself, Lord Rees of

Seeing a black hole in the blackness of space is not possible, but we can sometimes detect one when there is a companion star nearby. The black hole's immense gravity pulls matter from the star into a swirling accretion disc.

Ludlow, then known as Sir Martin Rees. Although Britain's top astronomer, he is rarely seen with his eye clamped to a telescope. Martin Rees is a theoretician, and an expert on black holes. He realized that if the cataclysmic events that produce GRBs are anything like black holes, there might be a way round the problem.

Rees knew that black holes do not emit energy in all directions, but in a narrow torch-beam emerging from the 'poles' of the black hole. Suppose that whatever caused GRBs did the same thing? The original calculations about the energy of GRBs had assumed that the radiation seen here on Earth was emitted in all directions and effectively filled the universe. That is why it had seemed so impossible to account for. The new idea was that the GRB we see is pretty much all the energy there is, that almost the entire energy

FACT FILE: BLACK HOLES

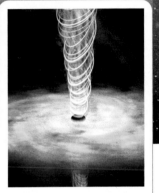

All stars may collapse when they run out of fuel, but only the bigger ones will form black holes. Once the star has collapsed to a 'singularity', a point that seems to contain all the mass of the star, the intense gravity can draw in other material from round about. This does not seem to happen from all around the black hole but from a disc of matter swirling around it. Black holes suck in matter, but the heat in the swirling disc also causes the hole to emit energy – as a narrow beam of X-rays at right angles to the plane of the disc, rather like torch-beams.

of the GRB has been directed our way in a narrow torch-beam. Martin Rees's idea was in one way a relief: it meant that GRBs don't break Einstein's most famous equation. But it also meant that GRBs must be far more common that anyone had thought. If we see only the gamma-ray beams that happen to be pointed at us, then there must be many more firing off in all the other directions. What are these things? And where are they coming from? Finding out has become one of the most urgent quests in space science today.

WORLDWIDE GAMMA RAY BURST ALERT

In 2006 the mission to track down the origin of GRBs got a powerful new tool. The Swift spacecraft is one of the fastest-reacting space telescopes ever built. It is at the centre of a worldwide rapid-reaction task force aiming to find out the truth about these very big bangs once and for all. GRBs are over incredibly quickly. Very short ones last only a third of a second or less. Longer ones average 30 seconds or so. How can you research something that will be over before you know about it?

> *'We get messages from the Swift satellite to our mobile phones'*
>
> Dr Kim Page,
> UK gamma ray burst
> advocate

Adam meeting Dr Julian Osborne, director of the UK Swift Headquarters at Leicester University, where a model of the sophisticated new telescope has pride of place.

Hubble snapped this image of the optical afterglow of a GRB. The object above it with the finger-like shapes is the originating galaxy.

Then there is another problem. One of the things astronomers most want to know about these events is how far away they are. This would normally be done with red-shift, analysing light to see how much redder it has become in its journey through the expanding universe. But you can't do red-shift with gamma rays – for that you would need light. Fortunately both problems can be solved with something called the 'optical counterpart'. But you still have to be quick.

Although whatever is emitting gamma rays does not itself shine in visible or any other light, the powerful burst of rays can make other things shine. As the GRB hits clouds of dust and gas, they glow, and this glow can be seen. What's more, the glow is peculiar and very characteristic because it starts off as energetic X-rays, then becomes ultraviolet, then visible light, then infrared, and eventually radio. This descent down the scale of electromagnetic radiation, like running your finger down the keys of a piano from right to left, can happen in minutes or hours – at most, days.

Although telescopes on Earth cannot see gamma rays, it would be wonderful if they could see the optical glow; perhaps then they could find out what is going on. But they would have to be alerted within minutes of a GRB going off. This is what Swift is designed to do, and if you are a scientist working on the project, the method is very simple and modern: the spacecraft will give you a call on your mobile phone.

The UK Swift headquarters at Leicester University is a standard office like thousands of others, but once every ten weeks it becomes the hub of a worldwide network of GRB detectives. The team at Leicester takes turns with other groups around the world to be 'burst advocates'. Their job is to be on call 24 hours a day in case Swift spots a GRB. Within a few minutes they must then assess how interesting a particular observation is, and if necessary alert the rest of the world so that the work of telescopes can be interrupted in the hope of capturing the fading glow.

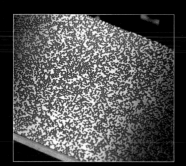

The BAT's ingenious metal shadow mask is made up of 54,000 lead tiles arranged in a random pattern. Each tile was placed by hand.

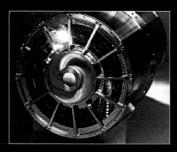

The XRT is actually a second-hand instrument. Both its mirrors and its CCD detector were designed for other missions but not used.

This is the UVOT being prepared in a clean room. It is tiny compared with most professional telescopes, but can pinpoint GRBs accurately.

SWIFT

Swift is a space observatory dedicated to the study of gamma ray bursts and their afterglows, which it can observe in the gamma ray, X-ray, ultraviolet and optical wavebands. The observatory comprises three main scientific instruments.

BURST ALERT TELESCOPE (BAT)

The BAT is Swift's gamma-ray detector. Gamma rays cannot be focused to form an image, so instead BAT uses shadows. Gamma rays coming in from space pass through a metal shadow mask. The mask is larger than the electronic detector 1 metre (3 feet) below, so rays coming in from different directions will cast a unique pattern onto the detector, allowing Swift to locate the direction of the burst.

X-RAY TELESCOPE (XRT)

X-rays cannot be focussed by conventional glass lenses or mirrors because they pass straight through without interacting. But X-rays do reflect off mirrors at very narrow angles, so the XRT uses a series of mirrors in concentric circles, each nearly but not quite parallel to the incoming X-rays.

ULTRAVIOLET AND OPTICAL TELESCOPE (UVOT)

UVOT is the most conventional instrument on board Swift, a 30-cm (12-inch) reflecting telescope that as its name suggests takes images in visible and ultraviolet light. If UVOT detects an optical glow from a GRB it can fix its position more accurately than either of the other instruments, which is important for other larger, Earth-based telescopes attempting to make observations of the afterglow.

SLEWING SYSTEM

Moving quickly to point XRT and UVOT in the direction of a GRB is called 'slewing', and Swift is designed to slew automatically within seconds of a GRB being identified by the on-board system. Slewing is done with reaction wheels – heavy wheels attached to electric motors so that when the wheel moves one way, the spacecraft goes the other.

The top of the Swift satellite, with the Burst Alert Telescope at the top, UVOT below it on the left, and the XRT on the right.

Swift turns to face a GRB. Its slewing system is powered entirely by electricity from its solar panels.

As a 'burst advocate', Dr Kim Page must take calls from Swift in the middle of the night if necessary.

DATA FILE: KIM'S PHONE

This is Kim's phone (right), showing the X-ray telescope (XRT) image of burst 061121. She got this automated call from the Swift spacecraft almost immediately after it had detected the burst, and the picture was enough to show that the burst had been recorded and located, and was probably worth following up. This burst would turn out to be the brightest ever recorded in the XRT.

When we visited Leicester, Dr Kim Page was to be the 'burst advocate' for the week. She warned us that while Swift usually finds one or more bursts each week, it was possible that there would be none at all during the week we were following her. She promised to phone us as soon as something happened. Then there was silence, an ominous six days of nothing happening at all. Finally, on the last day of her week in charge, the phone went, and she sounded pretty excited. Not only had Swift found a burst that afternoon, it was a really interesting one. We headed back to Leicester.

The burst was timed at 15.22 on 21 November 2006, and Swift's automated systems had reacted immediately, first slewing the spacecraft to point its telescopes in the right direction, and then phoning home. Because this burst was not in the middle of the night, Kim and her boss Dr Julian Osborne, who also got a call, arrived quickly in the Swift office ready to alert the world. Although Swift had reacted immediately, Kim could not hang about. Soon she also had an email from Swift, containing a record of what had happened, the first image, and a light curve. The light curve shows how the

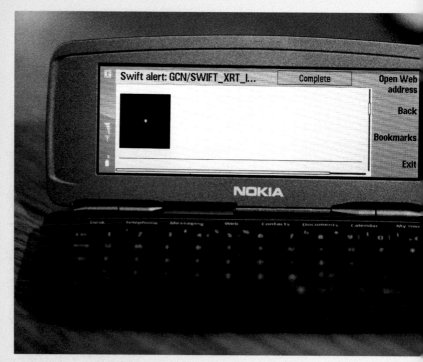

On 21 November 2006 the first image of burst 061121 is displayed on Kim's mobile.

brightness of the burst changes over time, and confirms that Swift is seeing the real thing. This one was particularly interesting. It seemed that Swift had triggered on a small 'pre-burst', a little blip before the real thing. This made it slew round into position and start taking images with its two telescopes – the first X-ray image was just 55 seconds after the alert. Then the really big burst happened 74 seconds later. This was the first time Swift had been in position and pointing all its instruments at a burst as it happened.

With an initial image and a light curve, Kim contacted the Swift teams around the globe in a telephone conference through a speaker phone on her desk. Within minutes, teams from Italy and the United States were all online, keen to know what had happened. They quickly agreed that Kim should send out an alert email to the rest of the astronomy world. This went at 15.39, just 20 minutes after the initial call from Swift.

Around the globe astronomers stop what they are doing and point their telescopes at the burst. Twenty seconds after Swift sees the burst, it is snapped by the Siding Spring Observatory, Australia. The astronomers at Siding Spring are ahead of the game, but in this case their telescope was responding automatically to the signal from the spacecraft, and snapped it before Kim's email went out. A few minutes later in Hawaii it is only twilight, but even so the 2-metre (6.5-foot) Faulkes North telescope picks up the automated message, sees the burst, and confirms its position. Also in Hawaii, the Keck Observatory, home to the world's largest telescope, manages to pick up the optical afterglow of the burst.

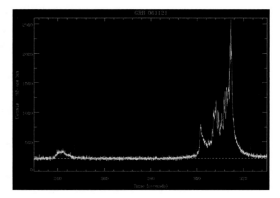

The sharp rise in the light curve indicates the GRB detected by Swift. This is all Kim needs to alert GRB astronomers worldwide.

Within minutes it has made a more detailed observation and has measured the red-shift of the glow, which suggests the burst happened 10 billion light years away. As the evening unfolds, more telescopes join in, and eventually there are further reports from Japan, the United States, Russia, and even other telescopes in space. It is one of the most 'followed' GRBs Swift has ever seen, partly because of the curious double burst, partly because it was very bright and so likely to have a good optical glow. Swift itself kept looking until 17 December, 26 days after the burst.

It is hard to grasp just how violent a GRB really is. There is more energy during the few seconds of a burst than our Sun emits in its entire life. Put another way, a burst is a million trillion times as bright as the Sun. Now that Kim, Julian and their colleagues round the world have shown that these huge bangs are happening a very long way away, there are ideas about what might be causing them. The longer bursts seem to be related to the deaths of stars, but on a much more massive scale than a typical supernova. So scientists call them hypernovae. In these immense stars the middle part – the core – collapses into a black hole, while the outer part explodes. Knowing how far away they happen deepened the mystery at first, because these giant dying stars seemed to be located at the places where stars were being born, the so-called star nurseries. Now it seems that their huge size means that these stars live out their whole lives, from birth to death, very quickly, so they die in the same place they were born. Short GRBs remain a mystery, but may be phenomena that occur when neutron stars collide.

Our knowledge of these biggest of bangs has a very short history. Forty years ago, no one imagined they could exist. Their accidental discovery by military satellites uncovered what many think is the greatest mystery in modern astronomy, explosions so vast they seem to defy the laws of physics. Now they are detected and measured almost daily, their stories tracked as they unfold by a network of dedicated scientists around the globe.

A massive star collapses in an explosion of gamma rays. Two jets erupt from the core at the speed of light. The jets form from the magnetic field surrounding a new black hole that is born from the star's collapse.

THE HEART OF THE MILKY WAY

Before there were cities with bright lights, the Milky Way was one of the most familiar features of the night sky. Old ideas about the universe and Earth's place in it could not adequately account for the distinctive band of white stretching across the sky. But in 1750 a mathematician from County Durham in the north of England published a remarkable book. Thomas Wright was a teacher and an amateur astronomer, but the title he chose for his lavishly illustrated work showed no modesty: *An Original Theory or New Hypothesis of the Universe*. Rather little is known about where he got his ideas, but he was one of the first to relate the night sky to the way we look at it.

Put simply, he thought the Milky Way was an optical illusion. He suggested that if Earth is located in a band of stars, and if we looked along the width of the band, then we would see very few stars. If, on the other hand, we looked along the length of the band, our line of sight would encounter many stars. So the Milky Way is not some special object, but simply how the stars of our system look in that very crowded direction.

Although his vision of Earth in a galaxy of stars is remarkable, Wright was wrong about the shape of the galaxy – he though that we were in a spherical 'shell' of stars (imagine an egg shell). Soon after his theory of the universe was published, the German philosopher Immanuel Kant and later William Herschel worked out that our galaxy is in the shape of a disc. But although these pioneers had ascertained the shape of the Milky Way, it took a further 250 years to find out what is at the heart of it.

The view of the Milky Way from the southern hemisphere is in towards the centre of our galaxy. But until recently you really couldn't see much there. Not only is this the direction most crowded with stars, but our view is obscured by clouds of gas and dust. So although we could see galaxies ten billion light years away, we were unable to see stars in the centre of our own galaxy. One of those leading the search for the heart of the Milky Way is Professor Reinhard Genzel, who runs the famous Max Planck Institute for Extraterrestrial Physics at Garching

DATA FILE: WRIGHT'S UNIVERSE

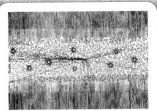

These are two of Thomas Wright's illustrations of the Milky Way, from his book *An Original Theory or New Hypothesis of the Universe* of 1750. He realized that if the stars are not distributed evenly, then you can account for the Milky Way by showing how we see more stars in one direction than in another. His account of how we see the sky was the first to suggest that we are part of a galaxy. In his book, Wright showed a universe filled with separate star systems like ours, a remarkable suggestion given that the belief that there was only one galaxy in the universe persisted until the 1920s. The illustration of an 'Eye of Providence' at the centre of each system was pure fantasy. However, new evidence suggests that there really is something lurking at the heart of our galaxy.

Reinhard Genzel has a great sense of humour, but wishes his organization had a different name, so that he didn't always have to deal with people's jokes about ET.

DATA FILE: SUPERMASSIVE BLACK HOLES

It seems certain that most galaxies have supermassive black holes at their centre. The image above shows a jet emitting from a black hole at the heart of galaxy M87. Possibly the black hole drives the galaxy in some way. What isn't clear is how it was formed, or which came first – the galaxy or the black hole. It is a question that may not be answered until we know how dark matter is involved in the formation of galaxies. When virtual universes are made (see chapter 1), the galaxies form around clumps of dark matter; it makes some sort of structure, invisible to us but having a huge effect on how the visible matter is gathered. We can now see the black hole in the heart of our galaxy, but it may be some time before we can tell its story.

in Germany. His quest for the centre of the galaxy has taken more than 15 years, but it has been worth it because he has uncovered something very big, and very violent.

The centre of the galaxy lies to the right of the constellation Sagittarius, the Archer. Disappointingly there is no visible object at the centre – or any sign of what lies within – as it is obscured by dark dust clouds. For some years radio astronomers had the clearest picture, and found that the centre is full of action – with curious filament-like structures, and in particular a huge source of radio waves they called Sagittarius A. Further investigation showed that within this big patch in the radio sky was a smaller but more intense radio source, called Sagittarius A*, or Sag A* for short (you say 'saj A star'). It is not incredibly bright, but Sag A* marks the true centre of the Milky Way, and this is where Reinhard focused his attention in the early 1990s. Using the world's largest telescopes that had a view of Sag A*, including the VLT, they took pictures of the stars near the centre in infrared light.

Although the pictures are impressive demonstrations of how infrared can be used to punch through gas and dust to see the centre of the galaxy, they did not contain anything very surprising. This particular hunt would need a great deal of patience. Reinhard kept photographing the same place, taking picture after picture. He has been doing it now for 16 years. His patience has finally paid off, however, because he is able to assemble his many pictures into a movie: each becomes a frame in a 16-year animation of what happens at the centre of the galaxy. The result is dramatic. Half a dozen giant stars loop around something invisible at the centre. It is like one of those fairground rides where you spin round slowly at first, and then whip round at high speed. The stars are obviously orbiting something, and it must be pretty big because they are travelling at 5 million kph (3 million mph). There was recently a celebration in Reinhard's department because the first star had been seen to perform a complete orbit, 16 years to go round the thing at the centre.

The invisible thing at the centre of the galaxy exerts an enormous gravitational pull. To make the stars dance as Reinhard has recorded needs something with a mass 2.6 million times that of our Sun. This huge beast is a black hole, and to make sure you know that it is not 'just' a standard black hole, formed from a star, they call it a supermassive black hole – formed from something much bigger. For a while all the evidence for the existence of the huge black hole was indirect: surely nothing but a supermassive black hole could account for what was happening? But more recently it has shown itself. It seems that some of the clouds of gas around the galactic centre are being fed into the hole. And as Reinhard puts it, the black hole 'snacks', and sometimes it belches. There is a flash of light as stuff disappears inside, never to be seen again.

This false-colour image from the Chandra X-ray Observatory shows the central region of the Milky Way. The bright, point-like source at the centre of the image was produced by a huge X-ray flare that occurred in the vicinity of the supermassive black hole.

147

DATA FILE:
THE ELECTRIC STORM
OF 1873

The following extracts are from the *Chicago Tribune* of 9 January 1873. 'The electric waves – so troublesome to the telegrapher – are variable in length, and vary from a second to one minute in duration. At times they act in conjunction with the battery current upon the wire, their united force grinding through the instruments with astonishing power, burning off the insulated covering from office wires, and melting the corners of brass machinery. All the marvellous power of lightning is displayed... In large telegraph-offices, where numerous wires centre to a common switch-board, the brass straps and faces are often illuminated by a constant succession of flashes interchanged between the several lines, which are harmless unless touched, and are beautifully attractive, especially at night... No warning ominously heralds the approach of these electric storms, and their departure is equally abrupt...'

OUR VIOLENT NEIGHBOUR

Our Sun is about 20 times too small to form even a normal black hole. But it will still do a fair bit of violence as it runs out of fuel, first swelling to become a red giant (and consuming Earth), then collapsing to form a bright white dwarf, and eventually fading away. The future of the Sun has been worked out by looking at other stars, seeing how they live and die, and then seeing where our star fits into the pattern. Seen this way, it is a pretty ordinary star. But for Professor Richard Harrison the Sun is entirely unique, because it is the only star we'll ever be able to examine in any detail.

The *Chicago Tribune* of 9 January 1873 carried an article reporting on strange and severe electrical storms that crippled the telegraph system (see box). An even greater electrical storm, in 1859, also caused fires. Much more recently, in March 1989, the Hydro-Quebec power grid in Canada went down for more than nine hours, causing considerable disruption and costing millions of dollars.

In 1873 the storms were perplexing because no one then understood that such dramatic electrical activity came from the Sun. At the Rutherford Appleton Laboratory in Oxfordshire, Richard Harrison monitors space weather. The space weather station has been cobbled together from bits of recycled kit, some of it dating from the 1960s when it was part of the Apollo programme's emergency landing site in Spain. Today, it displays data from the ACE satellite, downloaded through the huge receiver dish outside the laboratory. The hope is to be able to predict severe electrical activity soon enough to warn power and communication companies that their equipment is at risk. We are at far greater risk than the Victorians were: where they had a few telegraph lines, we have extensive power grids, satellites, and the Internet, all of which are vulnerable.

But what has the Sun to do with electricity? As with the other stories of violence in this chapter, the revelation of the true nature of the Sun came with technology that allowed us to see it in a new light. The most obvious thing about the Sun is how bright it is – so bright that you should never look at it directly. To do so with a telescope

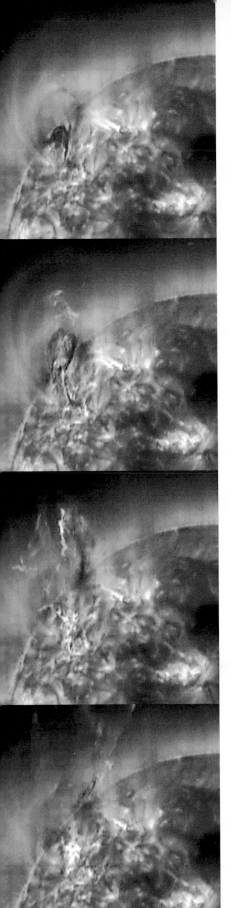

could cause blindness. The problem for Sun scientists is that the surface of the Sun, known as the photosphere, is so bright that it stops us seeing the interesting stuff, which is above the surface.

Other than the odd sunspot, the Sun in visible light looks smooth and untroubled. Slip a suitable filter over the front of your solar telescope, and a quite different picture emerges. In the light of glowing hydrogen, the surface now seems to boil. Step up to ultraviolet light, and something even more dramatic appears. Richard Harrison describes it as looking like a 'plate of writhing spaghetti', and that is a pretty good description. Erupting from the surface of the Sun are glowing hoops and loops, growing, twisting, sometimes breaking. Some of them grow to be huge. And sometimes when one snaps, a vast tower of flame erupts from the Sun and out into space. What is going on?

The first mystery is the extreme heat of the Sun's atmosphere. The surface of the Sun, the bit we see, is 'only' 6000°C (11,000°F). But the atmosphere above it can easily reach one million degrees. It is not clear why. To understand the loops, think magnets. You might have seen the trick of using iron filings to show up a magnetic field: put a magnet under a bit of paper, scatter iron filings on top, and they arrange themselves along the looping lines of the magnetic field. Incredibly, the same thing is happening in the Sun, but instead of iron filings these are electrically charged particles. Richard Harrison says that hot plasma is trying to burst out of the Sun, but is held in by magnetic 'elastic bands'. Sometimes these get twisted or break, and this is when the Sun can be dangerous.

For over ten years Richard's team at Rutherford Appleton has been part of the SOHO project. SOHO (Solar and Heliospheric Observatory) is a spacecraft that continually monitors the Sun, and has produced remarkable pictures of the Sun's violence. The Sun is not solid, but

The Sun's true violence can be appreciated in ultraviolet light. This sequence taken by SOHO over the course of an hour in February 2000 shows a coronal mass ejection. A CME blasts out millions of tonnes of particles at millions of kilometres an hour.

149

fluid, and the way the material of the Sun twists and rotates seems to control its cycle of violence. Every 11 years the storms and eruptions reach a maximum. The last maximum was witnessed by SOHO on Halloween 2003, when the flares and eruptions were even more violent than usual. Of particular interest are the coronal mass ejections, or CMEs. In length these are many times wider than the Sun, and can extend millions of kilometres into space. One CME can contain a thousand million tonnes of gas – nothing to the Sun, but potentially serious if it heads towards us.

Seeing the size and violence of the eruptions from the Sun, you appreciate Earth's protective magnetic field as never before. This acts as a shield, deflecting much of the 'solar wind' and some of the material from solar flares and CMEs. Without it, life would have been impossible on Earth. The radiation and stream of charged particles can damage cells and especially DNA, causing mutation and cancer.

Although SOHO has produced extraordinary pictures of our violent star, it has one severe limitation, as do Earth-based Sun telescopes: it can only look directly at the Sun. SOHO gets its images in two ways. To see atmosphere close to the surface of the Sun, it

uses filters to cut out the strong visible light and reveal the X-ray and ultraviolet. To see material being violently ejected from the Sun, it makes an artificial eclipse, blocking out the bright Sun with a circular disc called a coronagraph. This view reveals the flares and CMEs streaming 'sideways' into space. But these are not the ones heading our way. Hot gas heading down the line from the Sun to Earth is too dim to be seen against the bright atmosphere. The dream is to be able to stand off to the side, and take a picture across the Sun-Earth line. Now for the first time, this is going to be possible thanks to two new spacecraft.

Stereo was launched in late 2006 and released two craft that are travelling away from Earth in opposite directions. The two will operate together, taking simultaneous pictures of the Sun that can be combined to show it in 3D. They have cameras pointing directly at the Sun as SOHO has, but also cameras looking sideways for the gas reaching Earth, and these cameras have been provided by Richard Harrison's team. When we visited him in January 2007, the camera in one of the spacecraft had just been switched on for the first time. Even though it was not properly adjusted, it recorded a huge CME heading for Earth, and stretching over 10

Earth's magnetic field, called the magnetosphere, extends far out into space. The pressure of the solar wind squeezes it on the upstream side but extends it downstream into a long tail – the magnetotail.

The two Stereo spacecraft leave Earth. Inset: Stereo's first images of a CME heading directly at us – also a first for astronomy.

million kilometres (6 million miles). This had never been seen before. Once the cameras on both spacecraft are operating, Richard hopes to be able to tell the complete story of a CME from the moment it leaves the Sun to its arrival on Earth, travelling at speeds of up to 2000 kilometres (1200 miles) per second.

COLLISION COURSE: EARTH

We have seen how spectacularly the idea of a 'never-changing' universe has been thrown out by the evidence of a dynamic and often violent one. Most of the violence we have written about happens a long way from Earth. But we are by no means immune, as a headline in the *Guardian* newspaper of 7 December 2005 made clear: 'It's called Apophis. It's 390 m wide. And it could hit Earth in 31 years time.' 'It' is an asteroid, a piece of space rock, discovered in June 2004 by

the NASA Spaceguard Survey. 'Apophis', the name given to the asteroid, was in Egyptian mythology a demon that will plunge the world into eternal darkness. The width of the rock: '390 metres (1300 feet)' could 'affect thousands of square kilometres' should it strike Earth. So far, so bad. Our only hope seems to lie in that little word 'could' – and the efforts of several daring teams who are determined to do something about it.

When an asteroid entering the inner solar system is first spotted it is hard to pin down exactly where it is heading because it is usually quite distant from Earth. Hard, therefore, to tell whether it might become an NEO – a near earth object. Apophis was one such asteroid, and as it came

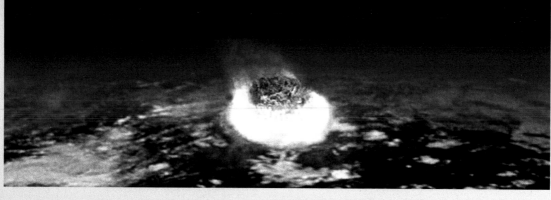

A digital artist's impression of a massive asteroid exploding in Earth's atmosphere.

closer, observers tried to fix its trajectory. By December 2004, there was serious concern. At one point it was given a 1 in 37 chance of hitting Earth on 13 April 2029. This is the highest risk of Earth impact ever given to an asteroid. When stories like this get into the newspapers, they are rarely taken seriously; the *Guardian* headline is clearly a bit of a joke. However, one look at the Moon might convince you to take the threat a little more seriously. Our satellite is pockmarked with the scars of thousands of impacts preserved on its pristine surface. But Earth is in the same bit of space, and, being larger, will have experienced even more impacts than the Moon. It is just that on Earth all evidence is erased by geology.

For the moment, the threat from Apophis has been downgraded. It will however pass incredibly close on the night of 13 April 2029, near enough to be visible to the naked eye from London. It will be ten times closer than the Moon. And in the closeness lies the danger. Earth's gravity will deflect Apophis in a way that can't be predicted, and might make it hit Earth when it returns in 2036, although that is not all that likely. Apophis will be on a collision course only if it is deflected to pass through a particular spot in space. The chance of it passing through this 600-metre (2000-foot) 'keyhole' has been estimated to be 1 in 5000. Although science fiction has dealt with killer asteroids, science has until now avoided the

challenge. But in the past couple of years, this has changed. NASA has studied Apophis, but decided it can wait until 2013 to make up its mind about what to do, because then there will be a chance to measure its orbit and work out if there really is a danger. The European Space Agency has decided to act – though not on Apophis. ESA will run a test mission to deflect a real asteroid, but one that poses no danger. The aim of the Don Quijote mission is to try out the technology so that we are ready when a real killer asteroid arrives. But how might you deflect an asteroid? The Don Quijote team has weighed up the options.

First, could an asteroid be blown up with a bomb? This is the Hollywood solution. Unfortunately this method might not save Earth: the danger is that the many pieces of the exploded asteroid would continue on a collision course. You might even increase the chance of collision by spreading the debris over more of the sky.

Second option: push the asteroid off course with a rocket. This is more controlled, and will not destroy the asteroid. The practical difficulty seems high, since the rocket and fuel must be transported to the asteroid and attached firmly. Physics says that even a small thrust will move a large asteroid. The worry is that we will not have sufficient warning of an asteroid impact for the time needed to make this method work.

Third option: paint the asteroid white. This is the simplest method, and relies on the pressure of sunlight shining on the white paint to move the asteroid off course. However, the force exerted by the sunlight would be tiny, and it seems unlikely that there would be enough time for this to be effective.

Fourth option: punch the asteroid off course with a heavy, very high-speed projectile. The asteroid would be deflected like a pool ball being hit by the cue ball. This method is relatively simple, and requires nothing to be attached to the asteroid. The main problem is that if you miss with your one attempt, Earth is doomed. On the other hand, something similar has already been done. On 4 July 2005 NASA smashed a washing-machine-sized probe into a comet. The Deep Impact mission was a great success, even though the targeting of the comet was described as like 'hitting a bullet with another bullet, which had been fired from a third bullet'.

Top, the Don Quijote mission orbiter, Sancho, records the impact with the asteroid. Middle, Sancho adjusts its trajectory. Bottom is the impactor, Hidalgo, heading towards the target.

The single punch is the method favoured by Don Quijote, and the mission is being designed in 2007. There will be two spacecraft: the orbiter, called Sancho, and the impactor, called Hidalgo. Sancho will stand by orbiting the asteroid, ready to take pictures while Hidalgo approaches at high speed. The big test will be whether Hidalgo can successfully hit the asteroid by adjusting its trajectory using a camera to track its target. Asteroid deflection must happen much too far away for navigational control from Earth – radio signals take too long to get there. So Sancho and Hidalgo will have to take care of themselves. Perhaps one day their sister mission will take care of us.

Of all the violent cosmic events described in this chapter, asteroid impact is the only one that has a real and immediate danger for our planet. Whether our species has a future may depend on whether we can use our technology to prevent an asteroid collision – something that only a few years ago humans would have been utterly powerless to prevent.

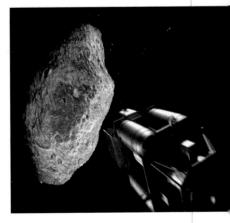

SAVING THE PLANET

Imagine Earth facing a mortal danger from outer space! If a large asteroid is heading towards us, could we deflect it? In our experiment on ice, a curling stone is standing in for the deadly asteroid. But which method is more likely to save Earth? The third scenario shown is the one that will be tested for real on a harmless asteroid in 2011 by ESA's mission, Don Quijote, which comprises two spacecraft. The first, Sancho, will use a single xenon ion engine to reach the asteroid and study it for several months. At the right moment the second spacecraft, Hidalgo, will 'charge' at the asteroid. Sancho will then check to see how much the asteroid's course has been affected by the impact.

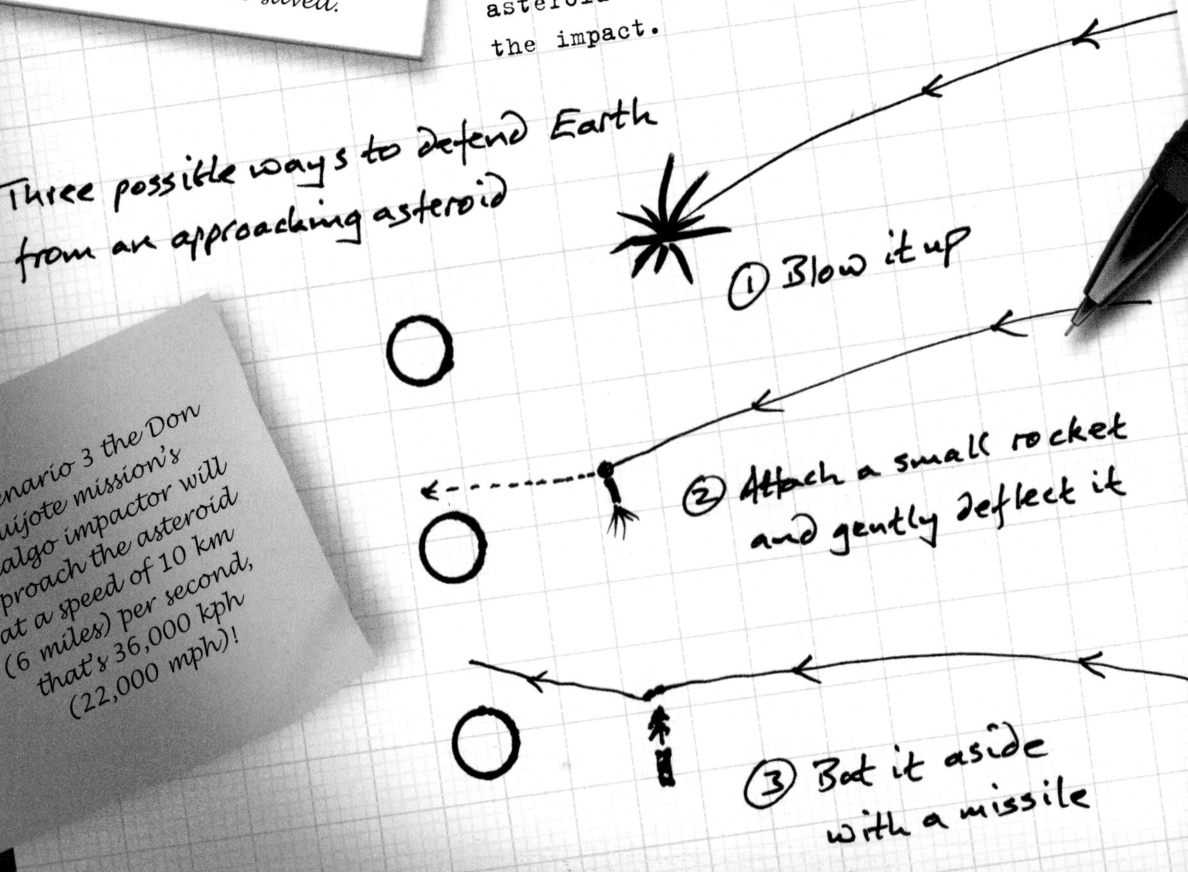

Presenter Maggie Aderin with Jason Dewhurst, a scientist bidding for the Don Quijote mission. This time it will be just a practice, so that we'll be ready should Earth need to be saved.

Three possible ways to defend Earth from an approaching asteroid

① Blow it up

② Attach a small rocket and gently deflect it

③ Bat it aside with a missile

In scenario 3 the Don Quijote mission's Hidalgo impactor will approach the asteroid at a speed of 10 km (6 miles) per second, that's 36,000 kph (22,000 mph)!

The Don Quijote mission is named after the fictional Spanish knight who charged at windmills thinking they were giants. Sancho was his squire who preferred to watch from a safe distance, which is what the Sancho probe will do.

Blowing up an asteroid heading for Earth is the solution favoured by Hollywood. But even if it were possible to destroy the asteroid, the pieces may continue towards Earth.

Another possible deflection strategy is to attach a rocket motor. Getting a rocket and its fuel deep into space is a challenge, and it will have to be attached by robot. What happens if (as they do) the asteroid tumbles and spins?

An ice hockey player shows how the Don Quijote mission might work. A single high-speed projectile will deflect the asteroid off course.

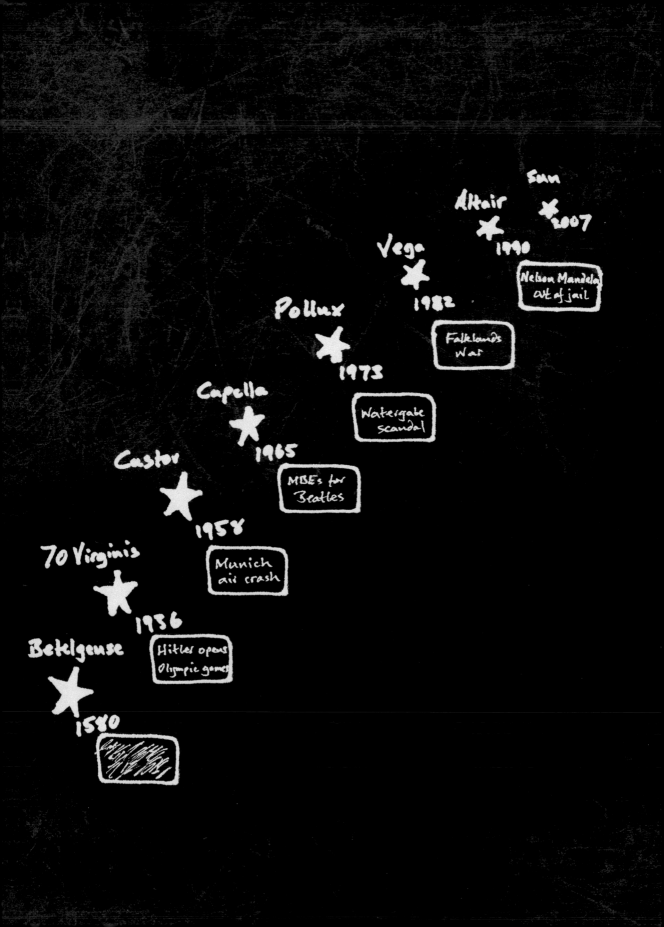

6 / ARE WE ALONE?

People are always fascinated by the question of whether or not we are alone in the universe. Human beings have been living and evolving on Earth for some four million years. Now, at long last, we have developed the technology to look elsewhere in the universe for other intelligent creatures – or at least for other living things. We live on a small rocky planet revolving around a fairly bog-standard star. The likelihood is that other similar planets revolve round other similar stars – and not only are there 100,000 million stars in our own galaxy, but there are 100,000 million other galaxies out there. Surely somewhere there may be life?

The question is more properly asked in two forms: is there any kind of life anywhere else in the universe apart from Earth? And is there intelligent life out there?

So far, even with all the science at our disposal, we don't know the answer to either question.

WHAT IS LIFE?

What do we mean by 'life'? Cats and dogs and people are alive, and so are trees and grass and geraniums, but also slugs and snails and mushrooms, not to mention plankton and bacteria. On Earth there are millions of different kinds of living things interacting with one another in countless ways. We know that life started with simple organisms of just a single cell, and gradually evolved over millions of years to the huge variety of life forms we see around us today.

If we look at the ways in which life began here on Earth, we can make a few intelligent guesses about what life could be like elsewhere in the universe.

How did life begin here? Where did those first simple cells come from? We still don't know for sure, but scientists have done a lot of thinking, speculating and experimenting. The answer lies primarily in chemistry. Biological cells are just bags of chemicals reacting together. Most of the chemical compounds involved are carbon compounds, generally known as organic compounds, and all the life we know about is based on the element carbon. The chances are that carbon may be the key to alien life, too.

In the 1920s Russian biochemist Aleksandr Oparin and British geneticist J B S Haldane suggested that simple organic compounds might have been formed from such gases as methane, ammonia and hydrogen – all present in the Earth's early atmosphere – in what Haldane called a 'prebiotic soup'. To react with each other these gases would need an energetic kick start – rather like the spark in the cylinder of a petrol engine. This might be a lightning strike, great heat, ultraviolet radiation from the Sun or perhaps even nuclear irradiation – and there is at least one group of natural, underground nuclear reactors here on Earth – in Gabon in West Africa.

In 1953 at the University of Chicago, American chemist Harold Urey and his student Stanley Miller decided to test the Oparin-Haldane theory of organic synthesis with the help of some laboratory lightning. Nowadays there are always thunderstorms raging

'There are infinite worlds both like and unlike this world of ours… We must believe that in all worlds there are living creatures and plants and other things that we see in this world'

Epicurus
Greek philosopher

somewhere on Earth, and there is a lightning strike every few seconds. If lightning was common four billion years ago, it would have been a good way of starting chemical reactions. The tremendous energy released by an electrical spark can rip stable molecules apart to make unstable fragments – highly reactive free radicals and ions – which can then attack nearby molecules. In a flask, Urey and Miller put methane, ammonia and hydrogen, along with some water. They kept the water simmering while passing electric sparks through the gases above to simulate lightning discharges. After a week of sparking, they found that the gases had reacted, and more than 10 per cent of the carbon atoms from the methane had become a mixture of other organic compounds, including amino acids. These are the building blocks of proteins, vital ingredients of both animals and plants.

The Urey-Miller experiment does not prove that life chemicals were made by lightning strikes. Indeed some scientists have argued that

The very earliest life forms on Earth may have been cyanobacteria – primitive single-celled organisms that flourished near the warm surfaces of the seas. Over time the bacteria formed colonies of mushroom-shaped mats known as stromatolites, which grew from the seabeds and solidified. Stromatolites (above) are still with us, absorbing carbon dioxide and releasing oxygen as waste, enriching Earth's atmosphere.

159

Hydrothermal vents, or 'black smokers', were seen for the first time only in 1977. The primitive bacteria that form the base of the food chain surrounding the vents provide tantalizing clues to what Earth's first life forms might have been like.

DATA FILE: EVOLVING LIFE

Until about 2000 million years ago the only living forms were simple bead-like microbes, such as cyanobacteria (above). Each one was no more than a single cell and reproduced by dividing in two. Over time, some microbes absorbed others and became larger cells, each with a cell nucleus. Eventually, these combined and 'specialized' to form tissues and organs in one many-celled body. This allowed complex life forms to flourish and evolve.

the early Earth's atmosphere was rich in carbon dioxide, in which case gases would not have reacted to produce amino acids. But the experiment does show that such a process is possible. And if it is possible on Earth, it is possible elsewhere in the universe, too.

For centuries, scientists thought that life, once it had taken hold on Earth, was sustained and nurtured by light from the Sun. All the plants and animals familiar to us depend on sunlight, for plants use light to drive the process of photosynthesis, turning carbon dioxide and water into carbohydrate and oxygen. Animals eat plants – or they eat other animals that eat plants – while breathing the oxygen, and they get their energy by using the oxygen to oxidize the carbohydrate. So at first sight it seems that sunlight is necessary for life.

A DISCOVERY IN THE DEEP

Recently, however, many surprising forms of life have been discovered in the darkness of the ocean floor – overturning the long-held view that all life is dependent on the Sun. We know less about the ocean floor than we do about the surface of the Moon, but a few brave scientists have crammed themselves into tiny titanium spheres and descended two or three kilometres into the pitch-black depths to investigate.

For thousands of kilometres along the length of the mid-Atlantic trench, and in other oceans too, the sea floor is continually spreading, peeling open by a couple of centimetres every year as Earth's tectonic plates drift apart. Up through the crack spews water superheated to about 400°C (750°F). It does not boil because of the enormous pressure at this depth. The water carries with it metallic

sulphides and other chemicals from the red-hot mantle below. Around these underwater volcanoes, or 'black smokers', is an astonishing abundance of living creatures, including tube-worms 2 metres (6 feet) long, and thousands of clams, mussels, crabs and shrimps, which crowd around spires of sulphides 10–20 metres (30–60 feet) high. These strange beings live without any light at all. They don't bother with photosynthesis; instead they bask in the heat of the water, and they use an entirely different sort of biochemistry to survive and grow. Some bacteria even seem to emerge from the vents in the superheated water, where no 'normal' living things should survive.

Could life on Earth have started with some curious life forms in the darkness of the ocean floor, using all the nutrients spraying out from the underworld? That is at least a possibility, and not necessarily less likely than another startling idea, that life did not originate on Earth at all, but arrived here from outer space.

Giant tubeworms surround the 'black smokers'. The worms have no mouths or guts. Instead they contain spongy brown tissue in which bacteria live. The bacteria get energy by oxidizing the hydrogen sulphide that bubbles out of the vent, reacting it with the oxygen dissolved in the water to produce sulphur, which then crystallizes out; it does not form shell, or anything obviously useful; it merely forms crystals of pure sulphur in the spongy tissue.

DID LIFE COME FROM OUTER SPACE?

We now know that even in the emptiness of space there are many kinds of molecules floating about, and some of them are complex organic compounds. Organic compounds have been found inside meteorites – the chunks of metal and rock that occasionally fall to Earth from space. There is even some evidence to suggest the existence of bacteria both in space and in meteorites. The idea that life might have started all over the universe by seeding from space is called panspermia (see box). Where did those organic compounds come from? One wild theory is they were left behind, either by mistake or on purpose, by alien creatures touring through the galaxy. Another possibility is that they were formed naturally from simple molecules, as in the Urey-Miller experiment, under the influence of ultraviolet light from the Sun – and presumably from other stars too. This theory is perhaps more probable on the basis of an old maxim called Occam's razor: that you should always seek the simplest explanation – because the simplest explanation is usually the right one – and therefore you should not assume the existence of aliens if you do not need to do so.

Bacteria have been found thriving in some of the coldest places on Earth, lending credence to the possibility that life could have been carried about the solar system inside the icy cores of comets. Seen here is the spectacular Comet Hale-Bopp, which appeared in our skies in 1995.

If organic compounds are drifting about in 'empty' space, then they might have floated down through our atmosphere – or fallen inside a meteorite – and started reacting in ponds and shallow seas when the conditions were favourable. Alternatively they could have been part of the original cloud of dust that was pulled together to create Earth 4500 million years ago. In that case they must have survived millions of years of Earth's turbulent early history before the climate settled down into a favourable state.

Various fragments of evidence support the idea that living things could migrate through space: high-altitude balloons have detected bacteria 30–40 km (18–25 miles) above Earth's surface, in concentrations apparently increasing with altitude, suggesting that the bacteria came from space – or that bacteria from Earth could drift into space. Also, sudden showers of red rain in the Indian state of Kerala in late summer 2001 contained red material that was originally thought to be dust, but appeared under the electron microscope to be living cells. Some scientists thought these were simple algal spores, but others claimed they contained no DNA, and must therefore be some form of extraterrestrial life.

The idea that bacteria could survive the vacuum of space may seem absurd, but they can be tough little things. In 1967 the unmanned American probe Surveyor 3 landed a television camera on the Moon; it was retrieved in 1969 by the astronauts of Apollo 12, and when examined was found to contain a little colony of a bacterium called *Streptococcus mitis*. These had survived on the Moon, in vacuum and

DATA FILE: PANSPERMIA

In 1996 a meteorite from Mars was found to contain what appeared to be fossilized bacteria (above). The find provoked renewed interest in panspermia. Panspermia is not a new idea; a similar concept was put forward by the Greek philosopher Anaxagoras 2400 years ago. But since the mid-1970s perhaps its most energetic champions have been British astronomer Sir Fred Hoyle and Sri Lankan mathematician Chandra Wickramasinghe. Hoyle died in 2001, but Wickramasinghe is today professor of applied mathematics and astronomy at the University of Cardiff. Both suggested not only that biological seeding from space kick-started life on Earth, but also that such seeding continues to this day, and may be responsible for some of our diseases and infections.

The bacteria that survived inside the Surveyor 3 camera (left) probably got there when someone sneezed during the pre-launch preparations.

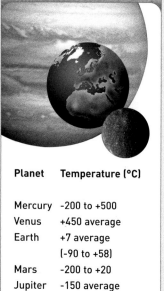

DATA FILE: PLANETARY TEMPERATURES

Planet	Temperature (°C)
Mercury	-200 to +500
Venus	+450 average
Earth	+7 average
	(-90 to +58)
Mars	-200 to +20
Jupiter	-150 average
Saturn	-180 average
Uranus	-200 average
Neptune	-220 average

'One explanation for the fact that we haven't found any life on Mars is just that we've been looking in the wrong places'

Charles Cockell
microbiologist

with extreme monthly temperature changes, for 31 months. So perhaps bacterial life could indeed have come from outer space.

There seem to be a number of ways life could have begun on Earth. What are the chances that life began on other planets, too?

LIFE IN OUR BACK YARD – THE SOLAR SYSTEM

For us it seems natural for a planet to be teeming with life, but are there other planets like Earth? The temperature on the surface of a planet is important, because any life even remotely like ours must surely be based on water. Human beings are more than half water, and all life on Earth depends on complicated chemical reactions going on in water solution. If you were to lose 15 per cent of the water in your body – perhaps as a result of severe diarrhoea or after serious burns – your biochemistry would pack up and you would probably die. Below 0°C (32°F) water turns to ice, and most living processes stop. Above about 70°C (160°F) water evaporates so fast that solutions rapidly become too concentrated to sustain complex reactions. Human beings have a thermostat system that keeps the blood halfway between these extremes, at 37°C (97°F). The other planets in our solar system have uncomfortable climates. The temperature on the surface of our sister planet, Venus, is about 450°C (840°F), so any water there would boil instantly. Mercury is even hotter during the day. Mars is cold; occasionally the temperature climbs to 0°C (32°F), but most of the time it's between -20°C (-4°F) and -100°C (-148°F). The outer planets are colder still.

So on the basis of temperature we should not expect to find any large-scale or complex life on the other planets in our solar system. However, there might well be an abundance of primitive, microbial life forms, perhaps dormant, that can survive in extreme environments. Just as the seeds of many plants can survive for years or even centuries without water, so bacteria and fungal spores can survive for immense periods of time in very hostile conditions – as did those bacteria on the Moon. We should not be blinkered by the living things we are familiar with. The weird creatures and plants flourishing by the 'black smokers' on the sea beds of Earth are called 'extremophiles', because they thrive in the extreme conditions of high temperature, pressure and darkness, which would be lethal to us frail humans. Those ocean-floor extremophiles are not unique: there are extremophiles in many other types of habitat, too.

THE MARTIAN POSSIBILITY

Even extremophiles could hardly cope with the conditions on the surface of Venus, although there are suggestions that some primitive life might survive in the clouds of sulphuric acid that permanently shroud the planet. On Mars, however, extremophiles might thrive, since conditions there are much less severe. Recent photographs from NASA's Mars Global Surveyor show what appear to be gullies full of ice or frost that have appeared in the last few years. This and other intriguing evidence suggests that there was and probably still is water on Mars – in the form of ice that is heated by geothermal processes and escapes as liquid water, only to be frozen again.

An image of water ice in a large crater at Mars's north pole was captured by the European Space Agency's Mars Express spacecraft. Subterranean water ice may well host extremophile life forms.

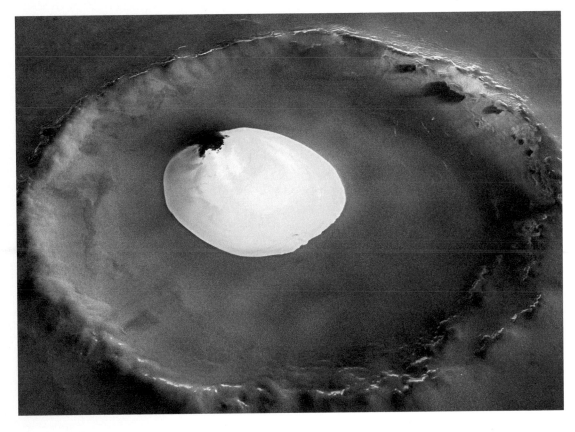

EXTREMOPHILES

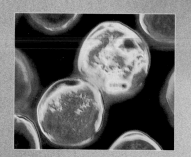

This is a micrograph image of **Archaeoglobus fulgidus,** *extremophile bacteria that live near deep-ocean hydrothermal vents.*

Extremophile bacteria have been found living in an amazing variety of habitats: in acid pools, in hot engine oil, in deep-sea temperatures of 170°C (340°F) surrounding 'black smokers', and in equally hot rocks far underground. Some are in sedimentary rocks, and feed off organic detritus that was deposited with them, but others are found in igneous rocks, where there is no natural food. The bacteria actually extract atoms of carbon from the rocks themselves, and use them to make methane.

Extremophile bacteria have been found inside floating icebergs such as these in Alaska.

Some bacteria have been discovered locked inside icebergs. Perennial springs bubble up through 400 metres (1300 feet) of permafrost at Axel Heiberg Island in the Canadian Arctic, bringing with the water long stringy bacteria and thin skins of biological material, even though the average temperature is a bitterly cold -17°C (1°F).

The bacteria may survive in these extreme places for thousands or even millions of years. Some scientists have even suggested that the mass of living things buried in the rocks is greater than that of all life on the surface of the Earth.

Since these extremophiles clearly stay alive well outside the 'normal' range of 0–70°C (32–160°F), we should also look for alien life well outside this range too, and also note that seeds and fungal spores are dormant forms of life that can survive extreme conditions for long periods of time and then spring back to life when conditions are favourable.

In the hot, acidic springs of Yellowstone Park, Wyoming, extremophile bacteria thrive in temperatures as hostile as 80°C (175°F).

An artist's impression shows how an ancient Mars covered in oceans of water may have looked.

> *'If you gave me a shovel and a microscope, in half an hour on Mars I could tell you whether there's any life or not'*
>
> Charles Cockell
> microbiologist

A billion or more years ago, Mars was almost certainly warmer and wetter than it is today and may have sustained life – and some bacteria or similar things might yet lurk underground, which is why the new explorer vehicles on the red planet are digging below the surface, looking for signs of life, or of past life. Such signs are more likely to buried since Mars now has little atmosphere, which means that the surface is constantly irradiated with ultraviolet light from the Sun, and a few million years on a sun bed would not be good for the health of anything. Ultraviolet light carries a lot of energy, so it is useful for kick-starting reactions in gases that aren't very reactive, but bad for living creatures, because those same reactions can be life-threatening.

During the 1960s NASA scientists were preparing to send astronauts to the Moon, which they finally accomplished in 1969, but meanwhile they wanted to set up the next major project for the 1970s. So they assembled a team at the Jet Propulsion Laboratory in Pasadena, California, to find ways of looking for life on Mars.

Among the people NASA invited to consult was the British scientist James Lovelock, who was rather scathing about the methods being proposed to hunt for the sorts of life common on Earth. He suggested instead that they should look for a large-scale reduction in entropy – in other words, evidence that something was making a mess. When his American colleagues looked blank, he explained that any living creature must take in some sort of nutrients or fuel and then excrete a degraded version; so it must need a waste-disposal system. On Earth animals and plants use the atmosphere and the oceans to get rid of their waste. One consequence of this is that our atmosphere contains both methane (produced by cows and sheep, at both ends) and oxygen (produced by green plants). These two gases should react together, and certainly do so over a long period of time; so the fact that they exist side by side in the atmosphere shows that something must be producing them continuously – in other words, their coexistence is evidence for life on Earth.

Lovelock suggested that NASA should examine the spectrum of the Martian atmosphere, which turned out to consist mainly of carbon dioxide, and showed no sign of such exotic mixtures as the atmosphere of Earth. Therefore, said Lovelock, there could be no life on Mars. This, however, was not the answer that NASA wanted to hear, so they fired him and carried on with what they had been doing – and 40 years later they are still looking for life on Mars. Today, however, the hunt is not for life as it was understood in the 1970s but for extremophiles beneath the surface – extremophiles that might not have exhaled strange gases into the thin atmosphere.

We know a good deal about the history of Mars. All planets are hot in the early stages, since the process of formation from clouds of dust generates heat. In addition there are many impacts with large chunks of rock, which also generate heat. When the initial

BIOG FILE : JAMES LOVELOCK

James Lovelock, born in 1919, is a rarity – he's an independent scientist. Most other scientists work in some sort of establishment because they need to be paid, and they need expensive equipment. Lovelock took a degree in chemistry and spent 20 years working in medical research, including investigating the causes of the common cold. In 1957 he invented the electron-capture detector, which turned out to be by far the best instrument for sniffing out the chemicals that were then polluting the environment. By 1965 he had broken away from all establishments and gone freelance, financing himself by selling his inventions. His extraordinary intuition allowed him to speculate about how Earth had supported life for so long, and to develop his theory that Earth is a self-regulating organism – the Gaia theory.

DATA FILE:
VIKING LANDERS

Several missions have visited Mars, starting with the Viking landers, which in 1976 tried to grow things they scooped from the surface by incubating lumps of soil with radioactive carbon monoxide and carbon dioxide. The purpose of this was to see whether any radioactive carbon was taken up and therefore whether any living organisms were in the Martian soil. The results were inconclusive; at first they seemed to be positive and to show there was life on the Red Planet, but later tests showed that the same results could come from inorganic material. Other Mars missions are described in chapter 4.

turbulence has settled down, the planet begins to cool, and Mars may well have had warm oceans for millions of years. However, because it is smaller than Earth it cooled more quickly, and to a lower temperature, mainly because it is further from the Sun, but also because it has less radioactive material in its core. Earth has enough uranium and other heavy elements to produce radioactive heating. Indeed this heating is what drives plate tectonics, the major geological movements of Earth's crust. On Mars, therefore, life could possibly have formed on the surface at an early stage, but would have been threatened when the atmosphere became too thin and the ultraviolet radiation too intense, or when the oceans evaporated. The living things might then have moved beneath the surface, and could still be there, lurking a few metres down.

Mars is close, and its temperature is reasonable, but we should not be short-sighted; some extremophiles are so hardy that they could exist on some of the planets beyond Mars, or on some of their moons. There could be living things on Europa, Ganymede and Callisto, three of the moons of Jupiter, or on Enceladus, a moon of Saturn's. They all have frozen oceans, and under the ice there may be deep liquid water with volcanic vents, and creatures perhaps similar to the giant tubeworms and shrimps in our own Atlantic. We may after all not be alone in the solar system – but the prospect of finding any large animals is slim indeed.

One reason to be pessimistic about the possibility of large animals elsewhere in the solar system is gravity. In order to be able to grow to a reasonable size, an animal needs water and an atmosphere. Planets much smaller than Earth have no atmosphere – the atmosphere of Mars is extremely thin – because there isn't enough gravity to retain the atmosphere, and prevent it from drifting off into space. Planets much bigger than Earth – the gas giants – have plenty of atmosphere, but there gravity is so powerful that any creature larger than a few millimetres high would be crushed flat by its own weight – if it could find a surface to stand on. An average person would weigh something like 200 kg (30 stone) on Jupiter, and human bones and muscles are not designed to take this force.

A digital artwork shows the Cassini spacecraft flying past Saturn's tiny moon Enceladus in 2005. Cassini photographed plumes of water vapour erupting from the surface, suggesting that liquid oceans, and potentially life, may lie below.

THE SEARCH FOR INTELLIGENT LIFE

We have to assume that there is no intelligent life elsewhere in our own solar system, even though we cannot altogether rule out a civilization of whales or dolphins, suitably adapted to low temperatures, in the seas below the ice of Europa.

What about planets in other solar systems? That is where more speculation begins, and people have argued the matter for centuries. Italian philosopher Giordano Bruno asserted that there are so many stars that some must have planets like Earth, and some of the planets must support life, but the Catholic Church disapproved of his cosmological ideas, and in 1600 he was burned at the stake in Rome. Similar speculation has intensified in modern times. In 1960 American astronomer Frank Drake, who was then a

'I am convinced there's intelligent life out there, and I'm also convinced that if we try hard enough, we will make contact'

Frank Drake,
SETI Director

171

BIOG FILE : FRANK DRAKE

Frank Drake was born in 1930. He studied electronics at Cornell University, but became interested in astronomy and in the possibility of life elsewhere in the universe. After a spell in the US Navy he went to Harvard and studied radio astronomy, and in 1958 landed a job at the National Radio Astronomy Observatory at Green Bank, West Virginia. He now works at the SETI Institute as Director of the Carl Sagan Center for the Study of Life in the Universe.

'Any number I give you has to be an enlightened guess, and my enlightened guess is 10,000; that is 10,000 detectable civilizations in our galaxy'

Frank Drake, SETI Director

radio astronomer at the US National Radio Astronomy Observatory at Green Bank, Vest Virginia, began to use the Tatel 25-metre (85-foot) radio telescope to listen for signals from a couple of nearby Sun-like stars. He called this Project Ozma, after the Queen of the Land of Oz (which, as everyone knows, is somewhere over the rainbow).

Project Ozma came to an end after a couple of months, but it gathered some publicity, and in the following year, 1961, Drake held a meeting to discuss the question of extraterrestrial life; there were just 12 people in the world interested enough to attend. In writing the agenda for the meeting, he developed what became known as 'the Drake equation'. This gives N, the number of intelligent civilizations in our galaxy capable of communicating with us, as:

$$N = R_* \times f_p \times n_e \times f_l \times f_i \times f_c \times L$$

- R_* is the rate of star formation in our galaxy;
- f_p is the fraction of stars that have planets;
- n_e is average number of planets that can potentially support life, for each star that has planets;
- f_l is the fraction of these that actually go on to develop life;
- f_i is the fraction of these that go on to develop intelligent life;
- f_c is the fraction of the above that are willing and able to communicate;
- L is the expected lifetime of such a civilization.

Unfortunately we don't know the values of most of the unknowns, but we have discovered a good deal about our galaxy in the last 40 years. First of all we now know for certain that there are many other planets out there (see chapter 3).

In 1961 Drake estimated that R_* is about 10 per year, $f_p = 0.5$, $n_e = 2$, $f_l = 1$, $f_i = 0.01$, $f_c = 0.01$, and $L = 10,000$ years. Using these values, N turns out to be 10; in other words he predicted that there should be ten other intelligent civilizations in our galaxy that are capable of communicating with us.

In 1984 the work of Drake and others in the United States and the Soviet Union led to the founding of SETI, the Search for Extraterrestrial Intelligence. SETI is a non-profit organization whose mission is 'to explore, understand and explain the origin, nature and prevalence of life in the universe'. The search goes on today. The University of California at Berkeley set up their own search system, called Serendip, at the Arecibo Telescope in Puerto Rico, and also took SETI to the masses by organizing a screensaver (SETI@home) that allowed people to do some of the tedious searching on their home computers – a neat example of massively distributed software. I even had it running on my computer for a couple of years.

So far the results have been negative. Apart from an early blip that turned out to have been produced by a secret U2 spy plane, no definite signal has been picked up in 40 years. There was another false alarm in 1967, when Jocelyn Bell, a young graduate student at Cambridge in England, was doing her PhD studying quasars, using a home-made radio telescope, but the source turned out to be a previously unknown object, a rapidly pulsating star, or pulsar, and her boss Tony Hewish shared a Nobel prize for the discovery.

Drake may have been too optimistic back in 1961. Some current estimates for the values of the unknowns in his equation give a value for N somewhere between 1 in 100 and 1 in a million. He himself, however, has become more optimistic still, and now believes there may be many more civilizations than the original 10 estimated in his famous equation.

THE 'RARE EARTH' HYPOTHESIS

Some people think there are thousands of intelligent civilizations in our galaxy; others think we are alone. How can their views be so different? The main reason is that there may be a huge gulf between simple life and intelligent life. Some scientists believe that although primitive life –

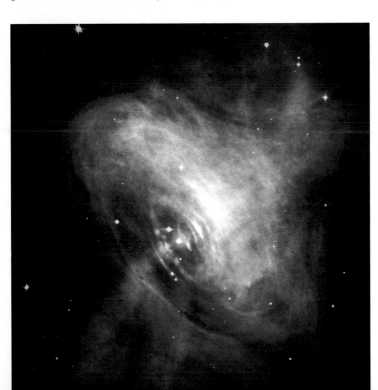

Radio astronomer Jocelyn Bell picked up a radio signal that repeated every 1.3 seconds with extraordinary regularity, and came from a distant point in space. Not knowing what it was, she called it 'LGM' (for 'Little Green Man'). The signal actually came from a pulsar – a rotating neutron star – in the heart of the Crab Nebula (left). Surrounding gases have been stirred up by the pulsar's powerful magnetic field and radiation.

173

bacteria, for example – may be fairly common in the cosmos, intelligent life may be exceedingly rare. This is known as the 'rare Earth' hypothesis, and is explored in detail by British scientists Peter Ward and Donald Brownlee in their book *Rare Earth*.

On Earth, life has been evolving for nearly four billion years, during which time vast numbers of species have slithered across the surface or swum through the oceans, yet only one species – *Homo sapiens* – has developed enough intelligence to be able to communicate through technology. Is our intelligence a natural or even inevitable outcome of evolution, or is it a one-off random phenomenon, never to be repeated? This is a vast question, and our entire outlook would be changed for ever if we did receive a definite signal from an alien civilization.

The American philosopher Daniel C Dennett points out that evolution is inevitable: the evolutionary algorithm, following from Charles Darwin's original ideas, says that if you have variation, selection and heredity, then you are bound to get evolution, or design out of chaos. However, no one can predict the outcome of such evolution, nor whether it is likely to lead to intelligence. After all, evolution a few hundred million years ago led to the dinosaurs, which dominated the planet for much longer than humans have, and yet the dinosaurs failed to acquire mathematics, music, technology or books. There may be thousands of planets crawling with dinosaur-like creatures, but we may never know.

WHERE DOES INTELLIGENCE COME FROM?

One theory is that to acquire intelligence, animals have to acquire language, which is difficult enough, but even before that they have to learn to imitate one another. Imagine a group of early hominids, of whom one person has learned over many years how to make a fire and keep it going, another has become a good cook and a third is a skilful hunter. Anyone who can learn to imitate all those skills is exceptionally well equipped for survival, and is therefore likely to attract many mates. As a result, imitators will have many children, and genes for imitation will spread through the population. Imitation does not necessarily depend on language, although learning a

The 305-metre (1000-foot) radio telescope at Arecibo, Puerto Rico, is the largest single-aperture telescope ever built. Set into a natural sink-hole in a dramatic landscape, the giant telescope has featured in several blockbuster films.

Learning to make tools was a crucial step in the development of our intelligence. It involves memory, planning and problem-solving.

language does require imitation. We find imitation easy; even young babies will smile back when you smile at them – and they can do this years before they have enough language to explain what they are doing. However, even though humans find imitation easy, no other animals can do it to anything like the same extent.

Some birds can copy other birds' songs, and even car alarms, chainsaws and other such noises, and it seems that a few apes can copy one or two simple actions. But in general humans are the only species capable of imitation. This may be the critical step; the sheer difficulty of imitation may be what makes us human, and may also make us unique in the universe. In other words the factor f_i in the Drake equation may be infinitely small. Unless we find another intelligent civilization we shall never know.

HOW SETI WORKS TODAY

SETI now employs about 120 scientists, who do all sorts of research, ranging from looking for other suitable planets to studying the habits and habitats of extremophiles in the hope that they may provide clues to the possible whereabouts of their cousins on other worlds. And above all SETI listens and watches for signals from space. The SETI headquarters are in Mountain View, California, just south of San Francisco.

Several telescopes have been involved in this work from time to time, including the Arecibo dish on Puerto Rico and the Green Bank Telescope (GBT). At 7000 tonnes the GBT is now the heaviest steerable object on Earth – and with a diameter of 100 metres (330 feet), larger than the other steerable telescope, Jodrell Bank near Manchester. When it is complete, however, the biggest listener will be the Allen Telescope Array (the ATA) at Hat Creek Radio Observatory, 500 km (300 miles) northeast of San Francisco.

The powerful new telescope will revolutionize SETI's work. With it, SETI's scientists hope to listen to signals from more than a million stars – quite a step forward from the two that Frank Drake started with.

The ATA dishes are arranged in a seemingly random pattern; in fact they are carefully placed so that every inter-dish distance is covered.

The dishes are brought in by road from Idaho and assembled on site in the Dish Assembly Tent. They are fitted with fabric shields to keep out the rain and snow.

ALLEN TELESCOPE ARRAY

Named after Microsoft's co-founder, Paul Allen, who provided several million dollars towards the project, the ATA will eventually comprise 350 separate 6-metre (20-foot) dishes, scattered over a square kilometre of California on the shoulder of Mount Shasta, the southernmost of the Cascade mountains. Swinging round in precise formation, the dancing dishes will all work together, which will make the ATA the most powerful radio telescope on Earth. There are several features of immense benefit to SETI.

First, the ATA will have an unusually wide field of view, so that it can listen to many star systems at the same time. For example, Arecibo can look in detail at an area of sky only about one-tenth the size of the Moon, while ATA will be able to observe an area about 22 times the size of the Moon.

Second, working in the microwave region of the spectrum, the ATA will listen simultaneously to about 10 million frequencies rather than having to choose one particular spot on the dial. SETI will have the use of ATA full time, 24 hours a day, instead of having to beg for a few hours at a time from one of the other telescopes. ATA will also be available for regular radio astronomy, but the search for extraterrestrials will continue round the clock.

Third, using the ATA, SETI should be able to look at more than 1000 stars in the first year alone, and then more and more each year, as computing power increases. In the last 40 years SETI scientists have looked at fewer than 1000 stars. The plan is to look at Sun-like stars, since they are most likely to have Earth-like planets in tow, starting with the nearest, which are likely to have the strongest signals. After choosing a particular patch of sky, the observers will be able to focus on two or three stars at the same time and will continue to observe them for perhaps ten minutes, before swinging the antennae round to the next target patch. This makes the assumption that an alien intelligence will be signalling continuously; otherwise there isn't much chance of picking up a signal in just a few minutes.

Seth Shostak, chief astronomer at SETI, presents the SETI weekly radio show **Are We Alone?** *He reckons we will have machine intelligence within 100 years, and that intelligent aliens probably already have, or even are, thinking machines.*

In the pod at the bottom of each dish there is an aluminium mirror to focus the incoming radio waves. There is also aluminium below the pod to shield the detector from ground radiation.

DATA FILE: ATA FREQUENCIES

Because the ATA's dishes are small compared with Arecibo they have a much wider field of view. Within that field, clever Fourier transform software will allow observers to look at several stars simultaneously. Computers will scan through all frequencies between 0.5 and 11 GHz. The observers will be looking for narrow-band signals, which must have come from a transmitter rather than a pulsar or a quasar. The signals will appear as diagonal lines on the screen.

Spectators at sporting fixtures often take flash photos from way up in the stands. Even though these flashes are feeble, they are easily visible from great distances away.

FLASH OF GENIUS

There is also an optical branch of SETI. In the early days, radio seemed the most obvious method of communication; the idea of anyone signalling with light seemed ridiculous. Now, however, intense and very short flashes can be produced by lasers, and these are an effective way of sending a signal in a particular direction. Even weak flashes can be detected – see how easily you can spot a camera flash from the opposite side of a stadium. An intense billionth-of-a-second flash could travel to us from many light years away, and still be recognizable as a signal. For that instant of time it would outshine the Sun by a factor of 1000, and would be easy to spot – by using a photomultiplier (an ultra-sensitive light detector) rather than the human

eye, which cannot respond to such a short flash of light. Signalling with light has to be highly directional, but you could convey information extremely quickly: an entire encyclopedia's-worth in a fraction of a second. Such a signal could come only from a civilization that was signalling specifically to us, rather than merely spraying out messages in all directions, but it would be easy to spot, and also cheap, since the total cost of the apparatus needed to receive the message would be a couple of thousand dollars, instead of the millions for a radio telescope.

Various observers in the United States and in Australia are already looking for such signals , and the work could be done by amateur astronomers, too. All you have to do is line up

178

your telescope on a likely star and observe continuously. If a signal were to come in, the intensity of the signal would briefly go up by a factor of about 1000, so your photomultiplier would bleep.

The SETI experts tend to observe one star for about ten minutes, and then switch to another. Just once they had a positive signal on all three detectors (the beam is split and fed into three detectors – in order to weed out false signals that are simply cosmic rays). Could it have been the real thing? An alien signalling? They have been back to observe that star many times since, without any luck. One drawback is that no one knows how long an alien would go on signalling to Earth. Perhaps it might signal only once in our direction, then move to another star, and go slowly through all the stars in the galaxy, always waiting for an answer.

So far optical SETI has looked at about 6000 stars, with only one possible result. Unfortunately this work is exceedingly tedious, and requires an observer, so it depends on the availability of supremely patient people.

SHOULD WE SEND OUR OWN SIGNALS?

This is a hotly debated question. On the one hand, if we are interested in finding out whether we are alone in the universe, then surely we should take a positive approach and ask questions – send signals rather than just listen. On the other hand, is it foolish or even dangerous so to advertise our presence? Might some super-advanced bunch of aliens be looking for a nice comfortable planet to invade?

In November 1974 the radio telescope at Arecibo in Puerto Rico was being upgraded, and the astronomers wanted to make an event of demonstrating the power of the 300-metre (1000-foot) dish. Frank Drake used the occasion to send a signal of immense power – a million times more powerful than a television transmission or the radio emissions from our Sun – towards M13, a cluster of 300,000 stars in the constellation Hercules. The British Astronomer Royal at the time, Martin Ryle, was furious. He said sending signals would be inviting attack from a civilization more powerful than our own.

DATA FILE: THE DRAKE SIGNAL

Drake's Areceibo signal (above) comprised a string of binary digits – 73 groups of 23 characters each – that could be interpreted as a sequence of messages: the numbers 1 to 10 (in binary 1, 10, 11, 100, 101, 110, 111, 1000, 1001, 1010), the atomic numbers of our 'life' elements (hydrogen, carbon, nitrogen, oxygen and phosphorus), representations of the structure of DNA, a matchstick person with height in terms of the wavelength of the transmission, the population of Earth, and a diagram of our solar system. This message was blasted out for three minutes, in the hope that some radio astronomer out there might be listening in 24,000 years' time – the length of time the signal will take to get to the M13 cluster.

IS ANYONE RECEIVING OUR SIGNALS?

The question of whether we should send signals is actually redundant. We have already been broadcasting radio signals for nearly 100 years and TV signals for about 70 years. The nearest stars outside our own solar system are about four light years away, which means that our signals take five years to reach them – because radio and TV signals travel at the same speed as light. However, at a distance 50 light years our signals have probably become too weak for any aliens to detect. What is more, less TV is being broadcast by transmitters; more and more is coming to us digitally. So as time goes by there is less for those aliens to watch and listen to. Conversely, this could mean that we are unlikely to be able to tune in to any alien broadcasts. Their signals will probably be too weak to detect, and any civilization more advanced than ours will probably have moved on from big transmitters anyway. Holding any sort of 'conversation' would be made very difficult by the time delay. To exchange a simple greeting with an alien on a planet in orbit around the nearest star, Proxima Centauri (4.2 light years away), would take over eight years.

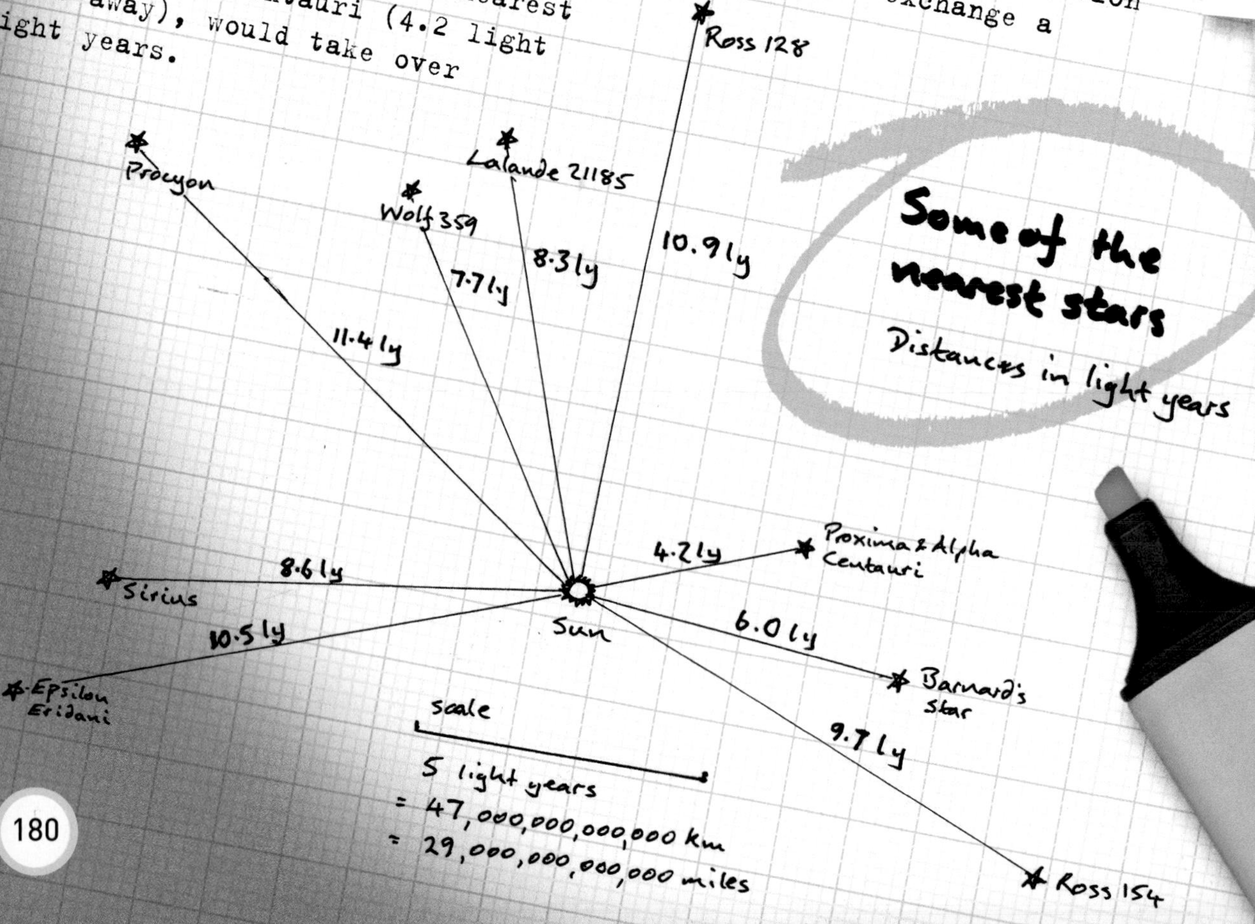

Ross 128

Procyon

Lalande 21185

Wolf 359

7.7 ly

8.3 ly

10.9 ly

11.4 ly

Some of the nearest stars

Distances in light years

4.2 ly — Proxima & Alpha Centauri

Sirius

8.6 ly

Sun

6.0 ly — Barnard's Star

10.5 ly

Epsilon Eridani

9.7 ly

scale

5 light years
= 47,000,000,000,000 km
= 29,000,000,000,000 miles

Ross 154

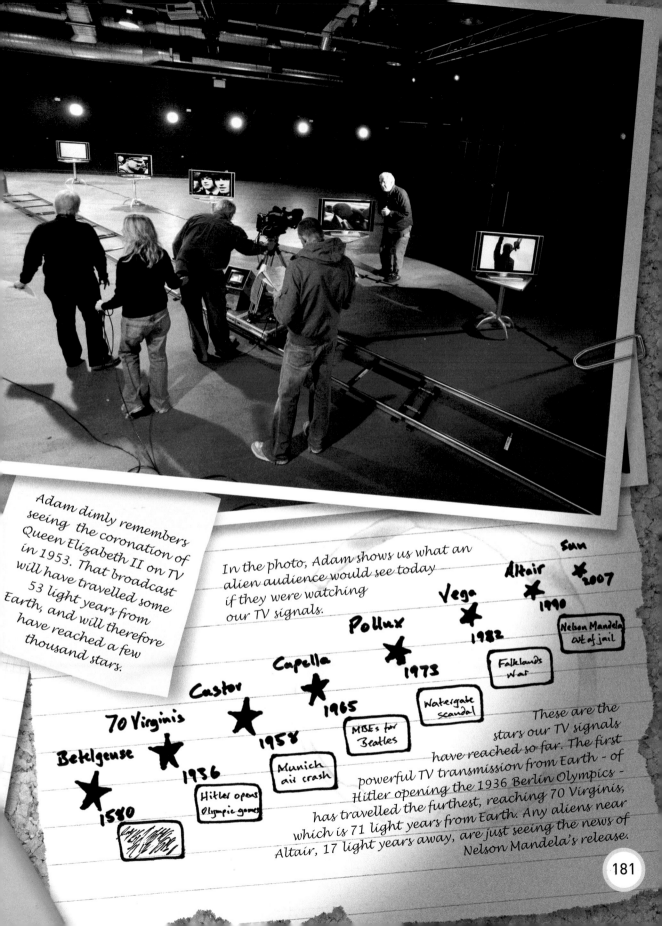

Adam dimly remembers seeing the coronation of Queen Elizabeth II on TV in 1953. That broadcast will have travelled some 53 light years from Earth, and will therefore have reached a few thousand stars.

In the photo, Adam shows us what an alien audience would see today if they were watching our TV signals.

Sun *2007

Altair 1990 | Nelson Mandela out of jail

Vega 1982

Pollux 1973 | Falklands War

Capella 1965 | Watergate scandal

Castor 1958 | MBEs for Beatles

70 Virginis

Betelgeuse 1936 | Hitler opens Olympic games

1580 | [illegible]

These are the stars our TV signals have reached so far. The first powerful TV transmission from Earth – of Hitler opening the 1936 Berlin Olympics – has travelled the furthest, reaching 70 Virginis, which is 71 light years from Earth. Any aliens near Altair, 17 light years away, are just seeing the news of Nelson Mandela's release.

Sending signals is unlikely to be dangerous. If aliens receiving Drake's message somewhere in M13 immediately jumped into their ship and set off to attack us, and even if they were able to travel at the speed of light, they could not reach us for 48,000 years. Since they are most unlikely to be able to get to more than, say, half the speed of light, we have some 75,000 years to prepare. And in any case, why should they bother?

In spite of these difficulties, the question of whether to send out signals is worth discussing. Do we want any other civilizations to know we're here? Are there really any dangers? Are there any benefits? Is it likely that any other civilization would be able to harm or help us? We have had the technology to send and receive signals for only about 50 years, so we are mere beginners. Any other civilization with radio telescopes is likely to be far more advanced. Therefore it is possible that we might be able to get guidance from them about how to proceed – how to advance our technology and avoid self-destruction. In other words we might learn something about possible futures.

WHAT SORT OF SIGNAL SHOULD WE SEND?

Pioneer 10 – a spacecraft launched in 1972 and now 8 billion miles away, heading out of the solar system – carries a plaque (designed by Frank Drake and Carl Sagan) with pictures of a man and a woman, while Voyager 2 – launched in 1977 and nearly as far away – carries a record of Beethoven, Bach, rock and roll and ethnic music, plus greetings in 50 human languages, which seems absurdly complex. Aliens are unlikely to understand any human language; to give them a challenge of 50 seems positively cruel. The record also carried sounds of rain, humans, cars and a kiss, which NASA insisted must be heterosexual.

The best kind of message is not easy to choose. There are two basic questions:
• How can we send something aliens can understand?
• What do we want to say?
Aliens may have language, but they certainly won't have any of our languages. Presumably if they have the technology to receive radio signals they must have some mathematics, and

The target for Drake's signal, the globular cluster M13, is in fact an unlikely home for life. Even though it comprises some 300,000 stars they probably do not have any friendly planets, since the conditions in the cluster must be hostile, with supernovae, neutron stars and lashings of dangerous radiation. What's more, by the time the message arrives, M13 will have moved on, revolving steadily around the centre of the galaxy, and there won't even be any stars in the target area.

The Pioneer 10 plaque shows Earth's position relative to the galactic centre. Travelling indefinitely through space, the craft will probably outlast Earth and the Sun.

mathematics must be the same throughout the universe, so mathematics may be a sound basis for communication. But what information do we want to send them? Since 2002 there has been a series of international workshops on interstellar message design, attended by psychologists, artists and philosophers. The 'art and science' of message composition involves creating messages that 'unfold and evolve' in response to the receiver, expressing the human sense of beauty, composing music inspired by the structure of DNA, and explaining the logic of altruism – all without words. Some members of the group are working on the type of language-free codes that might be used to convey these messages – and also on how to decode any message we receive. Indeed the first task must be to try to make sure that something we receive really is a message, rather than merely interference or random noise. Some even hope to develop software that would decode alien messages automatically. Fortunately we earthlings have some experience of code-breaking – as evidenced by cracking the German Enigma codes of the Second World War and by solving the sublimely subtle problem of Egyptian hieroglyphs with the help of the Rosetta Stone (see box).

DATA FILE: ROSETTA STONE

The stone was carved in honour of Ptolemy V in 196 BC, and found in 1799. The writing is in hieroglyphs, Demotic (a type of shorthand) and Greek. The chances were that the three scripts said the same thing; so the Greek, which scholars understood, was a translation of the hieroglyphs. But no one understood hieroglyphs – were they phonetic, or did the pictures represent words? Plus the language they represented had been dead for 1600 years. In 1814 English polymath Thomas Young guessed that one cartouche – an oval shape enclosing some hieroglyphs – contained the name Ptolemy. He was then able to list the sounds made by particular symbols. The code was finally cracked by French scholar J-F Champollion, who by 1822 had worked out that some symbols stood for sounds, others for whole words.

DATA FILE:
FIBONACCI NUMBERS

Fibonacci numbers are named after Leonardo of Pisa (1180–1250), who was known as Fibonacci. In mathematics, the Fibonacci numbers form a sequence that starts like this: 0, 1, 1, 2, 3, 5, 8, and so on; to continue, keep adding the last two numbers in the sequence. Remarkably, Fibonacci sequences appear in natural patterns, such as branching in trees, the curve of waves and the spiral-seed arrangements of pinecones. The sunflower above displays florets in spirals of 34 and 55 around the outside.

'It could be that this [message sent into space] becomes the only evidence for our existence 25,000 years from now'

Frank Drake, SETI Director

Even with hieroglyphs, however, the cryptanalysts were tackling another human language, so however unusual and weird it seemed, the thought processes and the words were likely to be similar. If a hand or a foot is mentioned, the meaning is obvious, but an alien might have five tentacles, and call them prodgers. How could we guess what was meant? We don't even know whether aliens would have vision – or hearing – in which case sending them pictures and music is fairly pointless. If we are to have a chance of understanding an alien message, what we would need is something like a cosmic Rosetta Stone. ETs are likely to have mathematics, and they are certain to have the same chemistry as us. One piece of common information, therefore, is the periodic table of the elements. Both sides would recognize that, and it might be possible to progress from there to exchange other information. Another possibility is to use mathematics itself, and convey something about aesthetics using Fibonacci numbers, the golden ratio, and the patterns in the spirals of seashells and galaxies, but where we go from there is not clear.

HAVE SIGNALS BEEN COVERED UP?

Some sceptics claim that signals have been received, but have been hushed up by governments or the military. This is most unlikely to be possible; such information leaks out too quickly. One signal picked up by Arecibo looked genuine, but after 12 hours of checking turned out to be a false alarm. Even within that time they had a phone call from a *New York Times* reporter, who wanted to know about the alien signal. Frank Drake and other SETI scientists are adamant that the very first thing they will do when they get a signal that appears to be genuine is to tell the world, starting with all their astronomical colleagues, because they would want as much information and corroboration as possible before they decide what action to take.

HOW MANY STARS?

Even in our own galaxy, the Milky Way, there are about a hundred billion stars. A reasonable fraction of them must surely be somewhat like our Sun, and so of the right sort of size and age to support an Earth-like planet. Unfortunately, the odds are rather stacked against another Earth.

This infrared image is a direct view into the centre of our galaxy. The density of light emanates from billions upon billions of stars. How many are anything like our Sun?

First, most of the nearby stars go around in pairs or clusters, and it seems unlikely that many such multiple stars could hold on to a planet, which in any case would be bombarded with radiation – and that would go for any planet near the centre of the galaxy, where all sorts of unpleasant neutron stars and supernovae blast exterminating radiation all over the place. The number of stars in quiet suburbs like ours is quite small.

Second, stars larger than the Sun have much shorter lifetimes; so even if one had an Earth-like planet in tow, it would not last long enough for our sort of complex life to evolve – since that has taken some four billion years.

Third, 95 per cent of nearby stars are much smaller than our Sun, which means that they produce much less light and heat; so an Earth-like planet would have to be uncomfortably close to the star, and would be in continual danger of getting pushed into an unstable orbit, or toasted by flares. Optimists point out, however, that such a planet would behave like our Moon. The Moon is tidally locked in position around the Earth so that we always see the same side of it; an Earth-like planet of a small star would be similarly locked, and the same face would always be towards the star. The nearside of the planet might be scorched, and the far side frozen, but in between there must be a temperate zone where life could thrive, even though there would be no seasons, nor even night and day. What's more, these small stars have much longer lifetimes than those like our Sun; so there would be much more time for the evolution of complex life.

We are about halfway out from the centre of our galaxy, on the edge of one of the spiral arms. Stars much further out would be no use for life, since they don't have enough heavy elements, or metals, which are vital for the right sort of habitat. Without metals there can be no metallic core to a planet and no magnetic field to protect from solar radiation. Second, complex living systems need small amounts of metals for their biochemistry. For example human blood contains haemoglobin, the chemical that carries oxygen around the body; in each haemoglobin molecule there is an atom of iron, which is what holds the oxygen. If there were no iron on Earth we would have no blood.

Scientists have worked out that the number and extent of 'habitable zones' is quite small. For example, if Earth were 5 per cent closer to the Sun, or 15 per cent further away, animals could not have evolved here. Bacteria, however, are tougher, and for them the habitable zone is wider.

Drake estimates that there may be 10,000 Earth-like planets in our galaxy, but there are several other factors that seem to have made Earth exceptional. First, we have Jupiter working like a bodyguard. The space in and around the solar system is full of debris – especially asteroids, which are lumps of rock orbiting our Sun, and comets, the dirty snowballs. Jupiter is so massive that its high gravitational field must have sucked in many asteroids and comets that would otherwise have smashed into Earth. We must thank Jupiter for staving off many life-threatening collisions. Jupiter also has an orbit that is almost circular, which may well have forced the other planets into near-circular orbits too. Most of the exoplanets

Comet Shoemaker-Levy 9 collided with Jupiter in 1994, leaving dark impact marks on the planet. Jupiter's gravity may have sucked in many projectiles that would otherwise have been on a collision course for Earth.

found so far have much more eccentric orbits. If Jupiter's orbit were as eccentric as theirs, Earth would almost certainly have been unstable, and would probably have been flicked out of the solar system altogether.

Second, we have one large Moon, which has played an important part in stabilizing the Earth's tilt and magnetic field. Without this stability the Earth might be tumbling uncomfortably in space, which would produce chaotic and rapidly varying climate, and make the long-term survival of life impossible.

Third, the Earth itself seems in a curious way to be self-regulating and continually maintains the conditions needed for life. This idea was introduced by James Lovelock in his book *Gaia*. He pointed out that several important factors have somehow remained favourable for several million years, which is at least surprising. For example the surface temperature of the Earth had stayed within the range necessary for liquid water to exist, even though the amount of heat Earth receives from the

Sun has increased by 30 per cent in the last three billion years. Ever since bacteria turned the atmosphere from a reducing mixture of hydrogen and methane to an oxidizing one of oxygen and nitrogen, the percentage of oxygen has remained between 16 per cent – the minimum necessary for mammals to breathe – and 25 per cent – above which forest fires would never go out. Lovelock argues that Earth's biosphere has developed feedback mechanisms to control such vital factors as temperature and oxygen concentration. Perhaps such a system could develop on another planet, but without it there would be little chance of conditions remaining stable long enough for intelligent life to evolve.

So, are we alone, or not? The chances of there being primitive life somewhere away from Earth seem to be quite high, and our hopes of finding it are growing steadily. Surely before long we will discover extremophiles snoozing away under some Martian rock. But intelligent life is more doubtful. There are pessimists who think that we earthlings are unique brainboxes in an otherwise unintelligent universe, and there are optimists who are convinced that the cosmos bristles with Mensa-worthy extraterrestrials. Frank Drake says, 'I am convinced there is intelligence out there. I am also convinced that if we try hard enough we will make contact.' Is he right? I don't know. We can only watch, listen, and wait.

DATA FILE: THE CARBON CYCLE

One of Earth's self-regulating mechanisms is the carbon-dioxide-rock cycle. This acts as a negative feedback loop, and so helps to control Earth's temperature. When the temperature is high there is violent weather, and therefore much weathering of rocks. This exposes calcium silicate, which reacts with carbon dioxide to make calcium carbonate, or limestone. CO_2 is a greenhouse gas; so when it is removed from the atmosphere the surface cools, and the amount of weathering decreases. But eventually the limestone will be heated up in the mantle, and the release of CO_2 from volcanoes means that the temperature will start to increase once more.

Seen from the Moon our beautiful planet does indeed seem rare. If we can survive our technological adolescence by not destroying it, we may yet live to find other worlds like ours – even ones supporting intelligent life.

GLOSSARY OF ABBREVIATIONS

ASI Agenzia Spaziale Italiana
ATA Allen Telescope Array (SETI)
AU Astronomical Unit; Earth's distance from the Sun
CERN Centre Européen Récherche Nucléaire
CMB Cosmic Microwave Background
CME Coronal Mass Ejection
COBE Cosmic Background Explorer
EGO European Gravitational Obvservatory
ESA European Space Administration
ESOC European Space Operations Centre
ET Extraterrestrial

EVA Extra-Vehicular Activity
GRB Gamma Ray Burst
HST Hubble Space Telescope
ISS International Space Station
JAXA Japan Aerospace Exploration Agency
LHC Large Hadron Collider
LIGO Laser Interferometer Gravitational-Wave Observatory
LTP Lunar Transient Phenomenon
MARS Mars Analog Research Station
MHU Mars Habitation Unit
NASA National Aeronautics and Space Administration
RTG Radioisotope Thermal Generator
SETI Search for Extraterrestrial

Intelligence
SOHO Solar and Heliospheric Observatory
STEREO Solar TErrestrial RElations Observatory
uv ultraviolet radiation
V-2 *Vergeltungswaffe-2* – Hitler's 'vengeance weapon'
VLT Very Large Telescope (Paranal, Chile)
WASP Wide-Angle Search for Planets
WCS Waste Collection System – the space lavatory
WFPC Wide Field and Planetary Camera
WMAP Wilkinson Microwave Anisotropy Probe

INDEX

ACKNOWLEDGEMENTS AND CREDITS

The authors would like to thank the many people who helped with this book, especially the Screenhouse team who produced the television programmes – Jess Baker, Nicola Dixon, Luke Donnellan, Amy Foster, Louise Say, Tracey Schawsmith, and Sarah Wilkinson; co-presenters Maggie Aderin and Janet Sumner who is also Head of Science Projects at Open University Broadcast Unit; David John, our meticulous editor at Tall Tree, who tried to keep us on course and on schedule; Martin Redfern at BBC Books who masterminded book production; Professor Paul Murdin, who did his best to pick up our worst howlers, and to all those who contributed to the television series in interviews: George Aldrich, Ben C Allanach, Francois Bouchet, Bob Chesson, Lars Lindberg Christensen, Charles Cockell, Dave Cullen, Joe Deeks, Jason Dewhurst, Michele Dougherty, Frank Drake, Helen Fraser, Carlos Frenk, Reinhard Genzel, Richard Harrison; Jennifer Heldman, Steven Jones, Verena Kain, Nick Kanas, Andy Kirk, Robert Landis, Mark Leese, Giovanni Losurdo, Nazzareno Mandolesi, Michel Mayor, Margarita Marinova, Vicki Meadows, Benoit Mours, John Murray, Don Nicholson, Terri Niles, Julian Osborne, Kim Page, Phil Plait, Thomas Paprafel, Don Pollacco, Didier Queloz, Alastair Reynolds, Ed Rhodes, Dave Rothery, Lynn J Rothschild, John Ruiz, Peter Schultz, Seth Shostak, Mark Sims, David Southwood, Ed Stone, Wes Trambe, Brian Turner, Laurence Tyler, Doug Vakoch, Pippa Wells, and Andrew Westphal. We would also like to thank the Open University, which commissioned the television series and has given us great support during its making. Special thanks are due to our excellent academic advisors, Dr David A Rothery and Dr Andrew Norton, who have provided invaluable advice.

Tall Tree Ltd would like to thank Chris Carpenter at Culham Science Centre, John Woodruff for proofreading, and Chris Bernstein for the index. BBC Books would like to thank the following individuals and organizations for providing photographs and for permission to reproduce copyright material.